NUCLEAR RUSSIA

Russian Shorts

Russian Shorts is a series of thought-provoking books published in a slim format. The Shorts books examine key concepts, personalities, and moments in Russian historical and cultural studies, encompassing its vast diversity from the origins of the Kievan state to Putin's Russia. Each book is intended for a broad range of readers, covers a side of Russian history and culture that has not been well-understood, and is meant to stimulate conversation.

Series Editors:

Eugene M. Avrutin, Professor of Modern European Jewish History, University of Illinois, USA

Stephen M. Norris, Professor of History, Miami University, USA

Editorial Board:

Edyta Bojanowska, Professor of Slavic Languages and Literatures, Yale University, USA

Ekaterina Boltunova, Associate Professor of History, Higher School of Economics, Russia

Eliot Borenstein, Professor of Russian and Slavic, New York University, USA

Melissa Caldwell, Professor of Anthropology, University of California Santa Cruz, USA

Choi Chatterjee, Professor of History, California State University, Los Angeles, USA

Robert Crews, Professor of History, Stanford University, USA

Dan Healey, Professor of Modern Russian History, University of Oxford, UK

Polly Jones, Associate Professor of Russian, University of Oxford, UK

Paul R. Josephson, Professor of History, Colby College, USA

Marlene Laruelle, Research Professor of International Affairs, George Washington University, USA

Marina Mogilner, Associate Professor, University of Illinois at Chicago, USA

Willard Sunderland, Henry R. Winkler Professor of Modern History, University of Cincinnati, USA

Published Titles

Pussy Riot: Speaking Punk to Power, Eliot Borenstein

Memory Politics and the Russian Civil War: Reds Versus Whites, Marlene Laruelle and Margarita Karnysheva

Russian Utopia: A Century of Revolutionary Possibilities, Mark Steinberg

Racism in Modern Russia, Eugene M. Avrutin

Meanwhile, In Russia: Russian Memes and Viral Video Culture, Eliot Borenstein

Ayn Rand and the Russian Intelligentsia, Derek Offord

The Multiethnic Soviet Union and its Demise, Brigid O'Keeffe

Nuclear Russia, Paul Josephson

Upcoming Titles

Art, History and the Making of Russian National Identity: Vasily Surkiov, Viktor Vasnetsov, and the Russia's History Painters, Stephen M. Norris

Russia and the Jewish Question: A Modern History, Robert Weinberg

The Soviet Gulag: History and Memory, Jeffrey S. Hardy

The Afterlife of the "Soviet Man": Rethinking Homo Sovieticus, Gulnaz Sharafutdinova

Russian Food since 1800: Empire at Table, Catriona Kelly

A Social History of the Russian Army, Roger R. Reese

NUCLEAR RUSSIA

THE ATOM IN RUSSIAN POLITICS AND CULTURE

Paul R. Josephson

BLOOMSBURY ACADEMIC

LONDON • NEW YORK • OXFORD • NEW DELHI • SYDNEY

BLOOMSBURY ACADEMIC
Bloomsbury Publishing Plc
50 Bedford Square, London, WC1B 3DP, UK
1385 Broadway, New York, NY 10018, USA
29 Earlsfort Terrace, Dublin 2, Ireland

BLOOMSBURY, BLOOMSBURY ACADEMIC and the Diana logo are
trademarks of Bloomsbury Publishing Plc

First published in Great Britain 2022
Reprinted in 2023

A catalogue record for this book is available from the British Library.

A catalog record for this book is available from the Library of Congress.

ISBN: PB: 978-1-3502-7255-2
         HB: 978-1-3502-7256-9
         ePDF: 978-1-3502-7257-6
         eBook: 978-1-3502-7258-3

Typeset by Integra Software Services Pvt. Ltd.
Printed and bound in Great Britain

To find out more about our authors and books visit www.bloomsbury.com
and sign up for our newsletters.

CONTENTS

ILLUSTRATIONS

TABLES

ABBREVIATIONS

BN	The Soviet Breeder Reactor
ICBM	Intercontinental Ballistic Missile
INF	Intermediate-Range Nuclear Forces Treaty (1987)
KIAE	Kurchatov Institute of Atomic Energy
kW	Kilowatt
LMFBR	Liquid Metal Fast Breeder Reactor
LTBT	Limited Test-Ban Treaty (1963)
MAD	Mutually Assured Destruction
MinsSredMash	Ministry of Medium Machine Building (the Soviet Nuclear Ministry)
MIRV	Multiple Independently Targetable Reentry Vehicle (multiple nuclear warheads on one missile)
MW	Megawatt
NPP	Nuclear Power Plant
NPT	Treaty on the Non-Proliferation of Nuclear Weapons (1970)
PWR	Pressurized Water Reactor
R and D	Research and Development
RBMK	Channel-graphite reactor (the Chernobyl type)
Rosatom	The inheritor of Minsredmash programs in Russia
SALT I	Strategic Arms Limitation Agreement (1972)
SALT II	Strategic Arms Limitation Agreement (1979)
SLBM	Submarine Launched Ballistic Missile

Abbreviations

START I	Strategic Arms Reduction Treaty (1991)
START II	Strategic Arms Reduction Treaty (1993)
UFTI	Ukrainian Physical Technical Institute
VVER	The Soviet/Russian PWR
WMD	Weapons of Mass Destruction
ZATO	Closed Administrative-Territorial Formation (closed military cities)

ACKNOWLEDGMENTS

The work leading to this book was funded in part by Colby College internal grants and assisted by discussions with students in my various courses. Erik Ananyan was an able research assistant. Funding from the National Science Foundation Grant 2020024, "A Global Nuclear Environmental History," was crucial in this work. Several colleagues over the years have helped me better to understand the evolution of the nuclear enterprise generally and the Soviet Russian case in particular. The anonymous reviewers for this book and the Russian Shorts series editors offered important suggestions and critical comments. Natalia Melnikova provided critical comments and corrections on an earlier draft of this book. Tatiana Kasperski has taught me a lot about the history, science, and politics of nuclear waste, about the Chernobyl disaster, about the public and technology, and about nuclear history across the globe. I dedicate this book to Tatiana with lifelong thanks.

INTRODUCTION

Nuclear Russia?

At the height of their power, the Soviets commanded 39,000 nuclear warheads, yet claimed to be servants of the "peaceful atom"—which they also pursued avidly. They built warheads and rockets, uranium enrichment facilities and reactors. They built the world's first reactor to produce power for the civilian grid. They embraced the atom as a sign that the advanced socialist system rivaled western capitalism, especially the United States. Physicists carried great authority in the USSR and today in Russia. Citizens were rarely permitted to protest against state programs, but in the case of nuclear power, at least until the Chernobyl disaster (1986–present), they, too, believed that the atom secured the nation's borders and provided cheap, safe and plentiful energy. This book examines both military and peaceful Soviet and post-Soviet nuclear programs for the long durée—before the Second World War, during the Cold War, and in Russia to the present.

Nuclear Russia covers much more than Russia. From the 1917 until 1991 the nuclear Soviet Union existed. After the breakup of the USSR, the nuclear nations of the former Soviet empire—Ukraine, Kazakhstan, Belarus, the Baltic states—continued major programs in nuclear research and development, in energy production, and in dealing with the legacy of Soviet nuclear waste, safety and other concerns. The atom in Ukraine has a long and difficult nuclear history. It was one of the founding nations of the world in nuclear physics in the 1920s and 1930s (as we shall see), contributed to the Soviet atomic bomb project after the Nazi retreat and defeat, had one-third of the Soviet nuclear arsenal, and, of course, is home to the Chernobyl disaster. To this day Ukraine produces over 50 percent of its electrical energy from nuclear power. Lithuania had two Chernobyl-type RMBK 1,500 MWe (megawatt electric) reactors on its territory, closing them both to accede to the European Union by 2010, although it continues to debate building another reactor of western origin. Kazakhstan was home to the Semipalatinsk nuclear test site of hundreds of weapons tests, the BN350 breeder reactor on the Caspian Sea, and had 1,400 nuclear warheads and undisclosed number of tactical nuclear weapons when the Soviet Union collapsed. Belarus, with its own extensive reactor development program, had eighty-one single warhead missiles stationed on its territory in 1991; all were transferred to Russia by 1996. In May 1992, Belarus acceded to the Nuclear Non-Proliferation Treaty (NPT). Today it is bringing on line a Russian-produced VVER1200 pressurized water reactor (PWR).

I will refer to these nations and their experiences as part of *Nuclear Russia*. But the focus remains Russia as the largest nuclear nation of the former Soviet Union. It often provided cadres and equipment to the other republics. It established and directed a national research and development program to all ends of the empire—to the Arctic northwest and the Far North and Far East, to Central and Western Siberia, and to Central Asia, especially Kazakhstan. The Kremlin actively instructed the direction of R and D that the other republics followed. And in the 2020s, one can argue that Russia is the most active nuclear nation in the world in terms of reactor construction at home and sales abroad, and, along with the United States, in maintaining and modernizing thousands of nuclear weapons.[1]

What is nuclear Russia? No different from many other large-scale technological systems crucial to the Communist Party and state in Russia for reasons of national security and economic growth, nuclear technology acquired features of scale, momentum and opaqueness about their true costs and benefits for the "toiling masses" whom they were meant to benefit. Because of the centrality of nuclear weapons to the postwar Soviet military and the importance of nuclear power to state economic plans, the Soviet nuclear enterprise never lacked for funding, institutional support, or first picks of the best students to staff their operations. The enterprise rapidly grew to hundreds of institutions, tens of thousands of employees, and dozens of entire cities devoted to the atom. When programs did not meet up to the promises of physicists and engineers—or there were accidents as in Kyshtym in 1957 and Chernobyl in 1986—the nuclear apparatus blamed local employees for their inattentiveness, mistakes or criminal behaviors, claimed that no such event would ever occur again, and confidently pushed forward to expand operations. For officials, scientists, and the public the "atom" became a panacea, capable of holding the west, particularly the United States, at bay, and able to generate copious amounts of energy in any environment, in any climate. Yet because of secrecy, the public was prevented from learning about the risks that scientists took to develop new nuclear applications, for example, in deploying mobile nuclear reactors or building the Chernobyl-type RBMK reactor with an inherent safety flaw and a lack of reactor containment. They did not know about haphazard disposal of nuclear waste; the dangers of radioactive fallout; serious accidents and exposures of the Soviet citizen to vast quantities of ionizing radiation; and so on. They learned of these things only in the Gorbachev era (1985–91), with the shock of these discoveries contributing to the loss of legitimacy of the Soviet regime. Nor were citizens fully aware of the on-again, off-again arms control efforts between the USSR and US that achieved some limitations in the kinds and numbers of weapons and delivery systems, as are discussed below in parallel with discussion of peaceful programs.

Hence nuclear Russia came to mean massive state support for the hubris of engineers as reflected in reactors, nuclear warheads, mobile power stations and a variety of other applications. While military applications were paramount in terms of budget and personnel, a public discourse that stressed the peaceful atom prevailed. This book shows how the peaceful and military atom developed hand in hand from the era of Joseph Stalin into those of Soviet leaders Nikita Khrushchev, Leonid Brezhnev and Mikhail Gorbachev. It discusses the challenges to the nuclear enterprise engendered by the economic and political collapse in Russia in the 1990s, and its stunning recovery, post-Chernobyl and post-Boris Yeltsin, during the presidency of Vladimir Putin.

Each of the following chapters is thematic and chronological, although in places I shall foreshadow important events and there is overlap of discussion of necessity since programs began and continued over the course of Russian history. Nuclear Bolshevism (Chapter 1) discusses the rise of nuclear science in Russia under the Bolsheviks, from the Revolution in 1917, until the Second World War. Nuclear physics found fertile ground to develop based on the efforts of a growing number of first-rate scientists and increasing support for science from the State. As was true throughout the world, even after the discovery of the neutron in 1932 nuclear physics grew in fits and starts of state support because of its high costs and the uncertainties, at least initially, over what applications might result from further study. Yet the science advanced rapidly, in part because of the compatibility of modern science with Soviet Marxism and its service to Bolshevik economic and political programs. Ukraine will be a major focus of this chapter, where the scientific enterprise grew against the backdrop—and horrors—of the Stalinist industrialization and collectivization campaigns.

Nuclear defense (Chapter 2) explores the genesis of the Soviet atomic bomb project, and how scientists convinced Stalin, his secret police chief Lavrenty Beria, and others to engage a crash program. The crash program gave rise to a series of secret institutes working on bombs and other nuclear applications that were supported by entire cities closed to the outside world, called ZATOs in the Russian acronym. Espionage played a role in the success of the Soviet atomic bomb project. But powerful domestic programs—and the ability of an authoritarian state to allocate all necessary support to them—were the ultimate keys to exploding the first atomic bomb in August 1949, only four years after the Americans had used nuclear weapons on Japan.

In 1953 US President Dwight D. Eisenhower gave his "Atoms for Peace" speech at the United Nations in the hopes of promoting international control of nuclear materials and know-how, and of advancing US business interests to sell peaceful technologies, especially from a nascent reactor industry, abroad. The Soviets were also pursuing peaceful applications and soon announced that the first reactor to produce electricity for the civilian power grid had come on line. The 1950s were a heady time for Soviet physicists. When Stalin died in 1953, Nikita Khrushchev assumed leadership of the USSR. He pushed a de-Stalinization campaign that led to greater cultural and political openness, and to more autonomy and power for scientists. Physicists in particular had great authority because of their contribution to the victory in the Second World War and their efforts to secure nuclear weapons during the unfolding Cold War. Building a kind of "atomic culture," they now pushed also a bold program of peaceful applications in industry, agriculture, and power generation. They claimed the "peaceful atom" for themselves and society. And they engaged the first steps toward arms control in view of growing fears over the danger to world public health over atmospheric testing, as Chapter 3 on nuclear peace explores.

If the vast majority of financial, institutional, and other aspects of the nuclear enterprise were directly connected to pursuing the arms race and achieving parity with the United States in powerful nuclear warheads and delivery systems (bombers, ICBMs, and submarines), the public face of the atom was its peaceful side in increasingly ambitious nuclear power programs. Chapter 4 on nuclear hubris focuses largely on the era of Leonid Brezhnev (1964–82) and the full flowering of plans to deploy power generating reactors in larger and larger units across the European USSR: pressurized water reactors (PWRs), channel graphite reactors (RBMKs) breeder reactors (LMFBRs), as well as smaller units, mobile reactors and floating reactors (icebreakers and reactors on barges). If plan targets were never met, and the industry encountered cost overruns and accidents, the scientists never lost their enthusiasm for nuclear power. Specialists fearlessly tested over 120 so-called peaceful nuclear explosions. They used nuclear power to tame the Arctic region—at great cost to indigenes. The nuclear industry even launched a factory, Atommash, to mass-produce standardized equipment for the nuclear industry. Granted, efforts at arms control continued, and Brezhnev signed agreements with Presidents Nixon, Ford and Carter. But arms control treaties ultimately did little to build down from Mutually Assured Destruction (MAD) between the two superpowers.

Mikhail Gorbachev intended to reform the Soviet economy and political system through perestroika and glasnost. He sincerely sought arms control deals in a series of meetings with US President Ronald Reagan. But the reforms unleashed powerful social and political forces of dissent, anger and frustration that he could not control, and as citizens learned about the lies they had been told about their history, and about the environmental and public health costs of the Soviet development model, a crisis in the legitimacy of the regime ensued. The Chernobyl disaster in April 1986 in many ways exacerbated this crisis by revealing vividly that the state was unprepared to help them in a technogenic emergency, indeed lied about what had happened and delayed evacuation, and the nation then suffered through a lengthy and costly cleanup that exposed still more people to mortal levels of radiation. Chernobyl triggered independence movements in several republics, and sparked environmental activism never seen before in the USSR. We can say that the disclosures about the lack of safety in the nuclear industry, on top of the Chernobyl accident itself, contributed to the disintegration of the USSR (Chapter 5).

But Russia is now living through a self-proclaimed "nuclear renaissance." If the atom fell on hard times in the 1990s under Boris Yeltsin, then it has found full support and full funding under President Vladimir Putin. The economic and political crises of the 1990s that saw many industries falter, and scientific research cease in many institutes, have past. The economy has rebounded on the sales of oil and gas, and Putin has determined to build a strong state on resource extraction, and to recreate Russia as a scientific superpower through achievements in space and nuclear power. In this environment, the nuclear industry has reappeared as Rosatom, a gigantic state corporation, with a determination to accelerate reactor construction at home and sales abroad. Opposition to nuclear power has been shut down by an increasingly authoritarian state, while Chernobyl, other accidents, and ongoing waste problems are acknowledged, but not instructive to temper the bold plans for the nuclear renaissance. The nuclear enterprise, under Putin, has also rekindled weapons development and deployment, if within the limitations of certain treaties, with new deployments of nuclear-powered submarines, cruise missiles and other weapons.

. The atom remains for Russia, in the twenty-first century, a crucial tool of domestic and foreign policy, and a central component of energy strategies, and industry leaders see no limits to its potential applications from already operating floating reactors to nuclear powered tankers and off-shore drilling platforms. In a word, from the 1930s under Stalin through 2020s the atom has been a central feature of Russian history: a tool of the economy, a cudgel in foreign policy, and a manifestation of a scientistic culture.

CHAPTER 1
NUCLEAR BOLSHEVISM

Tsar Nicholas II abdicated his throne on March 2, 1917, after the February Revolution, his corrupt and ineffective leadership no longer able to control the social turmoil that brought an end to the Romanov dynasty. The succeeding Provisional Government of moderate politicians, mostly of the middle class, attempted to rule Russia against a backdrop of growing anarchy in the countryside, a failed, ongoing war with the Kaiser's Germany, starvation and cold in St. Petersburg, and increasingly radical workers, soldiers and socialist political parties, including the Bolsheviks who were intent on violent seizure of power. By October 1917 the economy had collapsed, with the Bolsheviks seizing power. War Communism, a series of coercive measures to prop up industry by the Bolsheviks from the winter of 1918 to 1920 that included abolition of private property and the militarization of labor, accelerated economic free fall. Many members of the intelligentsia emigrated; others died of starvation, including seven of forty-two members of the prestigious Academy of Sciences; still others saw their livelihoods—and lives—expropriated by representatives of the Bolsheviks and criminals.

Yet out of this horrific violence, economic despair, and political turmoil arose a new, Soviet scientific establishment capable of participating in another revolution—a revolution in science marked by the rise of relativity theory, quantum mechanics, and nuclear physics. Although a number of analysts have suggested that the USSR built its nuclear bombs on the foundation of espionage and industrial theft, the Soviet Union was well-prepared to be a major player in nuclear age—and the Cold War arms race—because of institutional, scientific, and political developments of the 1920s that supported the creation of a modern scientific establishment. During that time, the

Bolsheviks and scientists reached accommodation on funding new facilities and expanding the research and development enterprise far beyond the level of Tsarist support. They saw to the training of increasing numbers of young, capable researchers, including women, in universities previously not open to many citizens. Before the rise of Stalin, when party officials were not as directly concerned about the political views of scientists as they would become, they facilitated reestablishment of scientific ties with the west, and they granted a fair degree of autonomy to scientists in running research programs. All of this ensured that physicists in Leningrad, Moscow and Kharkiv, Ukraine, a number of whom had studied in Germany and England, were prepared to join the 1932 *annus mirabilis* in physics when John Chadwick discovered the neutron, Carl David Anderson identified the positron and the first artificial disintegration was realized with a particle accelerator by John Cockcroft and Ernest Walton; these discoveries were followed rapidly by that of artificial radioactivity by Frédéric Joliot and Irène Curie, and of nuclear fission by Otto Hahn, Fritz Strassman and Lise Meitner in 1938. Soviet physicists were a part of these revolutionary discoveries.

There was a kind of strong compatibility between Marxism and the sciences with such leading Bolsheviks as Vladimir Lenin and Lev Trotsky asserting that scientists were natural materialists and could be relied upon to contribute to the building of the socialist state through their research. That research would give rise to productivity increases in industry, agriculture, and electricity production. They believed, similarly, in the universalism of science and technology, that is, that they must acquire and master the leading technologies in the world of the capitalist powers and bring them into perfection under socialism. The Bolsheviks sought to build their own technological systems outright—factories, components, and so on—but were also willing to buy western technology or engage in industrial espionage. As for science, already from 1918, the Bolsheviks supported the efforts of scientists to acquire journals, reagents, and equipment from the west, including through travel grants, and also the efforts of national organizations of physicists, geographers, geologists, biologists and botanists, and many other groups to hold conferences and carry out expeditions at home. This internationalization of science that lasted until the early 1930s was crucial to the history of Soviet nuclear physics, especially through the activities of a number of physicists who had studied in Cambridge, England, one of the leading centers for the study of the atom, and in particular those at a newly founded institute in an agricultural region of Ukraine.

In 1928 promising young physicists from Leningrad and other cities were ordered to Kharkiv to establish the Ukrainian Physical Technical Institute (UFTI) under the umbrella of the Commissariat of Heavy Industry.[1] This forced transfer of science and industry from the industrial heartland into other regions of the empire was part of the effort to harness modern engineering to Bolshevism. When Chadwick discovered the neutron in 1932 at Cavendish Laboratory where several UFTI scientists had studied, the Soviet physicists had already embarked on parallel research on the atomic nucleus at home, an effort that led to their institute eventually becoming one of the central laboratories of the atomic bomb effort.

The path to the nucleus opened in eastern Ukraine with great promise. A score of young, talented physicists gathered in Kharkiv, then the capital of Ukraine, to open a new institute connected with burgeoning machine-building, electrification and communications industries. The institute reflected the plan of Abram Ioffe, himself a specialist in solid state physics and the unofficial dean of the Soviet physics establishment, head of the Leningrad Physical Technical Institute (LFTI), to establish a network of such physical-technical centers around the country—in Tomsk, Sverdlovsk, Dnepropetrovsk, and Kharkiv—to provide cutting-edge science for Stalin's industrialization effort. The idea was that using core LFTI personnel as the kernel of a new research center, they would attract promising young local talent to a series of physical technical institutes established throughout the country. In each case, these institutes would support new or planned industrial centers. Kharkiv's location near the Donets Basin and along trade routes between Russia, the Black Sea, Central Ukraine, and the Caucasus were crucial to its growth. UFTI was founded in October 1928, although the institute did not open its doors officially until two years later.

UFTI stood out among the new institutes for the quality of its personnel who included specialists in low temperature and theoretical physics, a future Nobel laureate, Lev Landau, Alexander Leipunsky, one of the founders of the breeder reactor program of the USSR, and, for a brief spell, Igor Kurchatov, the father of the atomic bomb project. Kurchatov worked on a proton accelerator and in 1933 headed the organizing committee of the First All-Union Conference on the Atomic Nucleus held in Leningrad. (This was followed by nuclear physics conferences in Moscow in 1935, Leningrad in 1938, Kharkiv in 1939, and Moscow in 1940.) Kurchatov was later appointed director of the Soviet atomic bomb project, working with secret police chief Lavrenty Beria. Kirill Sinelnikov, the brother-in-law of Kurchatov, low-temperature specialist Lev Shubnikov, theoreticians the Lifshits brothers and others pushed the institute into major discoveries in a variety of fields. These dedicated physicists, some of whom admittedly did not wish to be transferred to Ukraine which they considered to be provincial in culture and behind the times in physics, instead quickly gathered a critical mass of scholars.

As a relatively new discipline in the USSR—chemistry tended to be more developed—physics provided opportunities to advance among young specialists who had been excluded from the sciences by a variety of Tsarist educational prohibitions. Jews in particular, but also Armenians, Georgians, and scientists from the Baltic states benefitted from the Soviet system, entering physics (and the other sciences) in numbers greater than their share of the population would indicate. Russians, of course, predominated by virtue of a larger share of the population, as did Ukrainians. Finally, there were many more men than women in the sciences, especially at level of laboratory or institute directors. In this case, UFTI had a large share of Jewish specialists.

UFTI was not a scientific colony of Moscow and Leningrad. Although the institute's basic research program was tied to Kharkiv's intended status as a big machine-building center, the level and quality of physicists, and the panache and originality of their research were world-class from the very start. Sinelnikov, Anton Valter, Landau, Obreimov, Shubnikov, and Leipunsky embarked on cutting-edge research. Niels Bohr, Paul Dirac, Paul Langevin, P. M. S. Blackett, Robert Van de Graaff, Paul Ehrenfest, and Boris Podolsky visited. The foreign physicists Alexander Weissberg, Fritz Houtermanns, and Laszlo Tisza joined the institute staff (the first two would experience the purges of the Great Terror firsthand). And UFTI physicists became pioneers in theoretical physics, in nuclear physics, in fusion, in the design of the Romashka nuclear battery, and in linear accelerators. They published extensively, won Lenin and Stalin prizes, and the institute earned a coveted Order of Lenin. International in their backgrounds, UFTI physicists worked to keep their institute open to foreign scholars who came to visit, lecture, and stay. Because few European and North American specialists read Russian, let alone Ukrainian, they established a German language journal to ensure their work came to the attention of colleagues abroad, *Physikalische Zeitschrift der Sowjet Union,* from 1930 to 1937, when it was closed down by the authorities, ostensibly to protect them from revealing state secrets.

Experiment and theory went hand in hand in Kharkiv. Like their colleagues in Cambridge, England, Rome, Italy, and Pasadena and Berkeley, California, Sinelnikov, and others pushed accelerator technology to penetrate deeper into atomic structure, at first using charged ions, and then, following the work of Enrico Fermi in Rome, irradiating various elements with neutrons. These elements absorbed neutrons, were transformed into radioactive isotopes and gained a unit of atomic weight, then through beta decay became a stable isotope. Lithium became beryllium, beryllium—boron, and so on. UFTI specialists built several of the world's first electrostatic and other particle accelerators. In 1932 UFTI's research plans included a new theme, "the study of the atomic nucleus with the help of collisions of fast particles." In 1933 the institute added the theme "research on the neutron—a new kind of matter." As a source, they bombarded beryllium with artificially produced alpha particles or deuterons. In 1935, Sinelnikov and his colleagues published two works on neutron absorption of neutrons, while Leipunsky, who studied in Cambridge, England, led a group investigating the scattering and capture of neutrons in a large group of elements. (After the Second World War

Leipunsky directed the Soviet breeder reactor program that is now centered at the Beloiarsk station in the Urals.)

But these achievements occurred against the backdrop of the rise of Stalin to party supremacy. In this new stage of nuclear Bolshevism, Stalinist Five-Year Plans for rapid industrialization and collectivization of agriculture were accompanied by cultural revolution in art, literature, film, higher education, and science. The Stalinists put pressure on scientists to conform to the dictates of the state for industrialization and to embrace an ideology of proletarian science. According to the adherents of proletarian science, scientists were suspect, either by virtue of their inherent technocratic tendencies or because, as remnants of the Tsarist era, they were hostile to socialism. Science had to serve the immediate and practical needs of the economy. The imposition of Stalinist science policies involved the centralization of policy making in such Moscow-based government, party and scientific bureaucracies as the Ministry of Heavy Industry. Centralization enabled officials to enforce an emphasis on applied science at the expense of basic research. In general, the five-year plans indicated that the periphery—and in the case of nuclear physics, Ukrainian physics—would serve the center. At the same time, in a fight against so-called bourgeois nationalism, the Party attacked Ukrainian intellectuals, arresting and purging hundreds of them in the attempt to eradicate all vestiges of Ukrainian nationalism in culture.

Stalin's programs led to immediate and dark changes in the daily life of Kharkiv. The huge new Kharkiv Tractor Factory (1931), the Kharkiv Machine-Tool Plant (1933), and the Kharkiv Turbine Plant (1934) were brought on line, and were quickly spitting out motors, generators, and other such equipment that supported electrification and collectivization throughout the nation. Urbanization accelerated with thousands of workers—many of them peasants fleeing the countryside—flooding the city. Forced collectivization led to suffering in the cities and villages. During collectivization and "de-kulakization" millions of peasants were forced to join collective farms, millions were exiled to Siberia and Arctic regions, and others sought refuge in cities. To feed the cities, the Bolsheviks requisitioned grain. Kharkiv swelled from roughly a half million people to 833,000 in 1939 as starving peasants flooded it and other cities to escape the Stalinist Holodomor (famine) in the countryside. When Stalin was informed of a spreading famine he ordered Ukraine's borders closed by police and soldiers. The Kharkiv region was one of the worst hit.[2] Just outside the institute's walls peasants who had left the countryside for cities in search of food died on the streets in view of the scientists; overall three million people likely died of starvation.

Figure 1 A famine struck in Ukraine in the early 1930s as a result of collectivization of agriculture and an attack on the peasantry. Millions died. This Ukrainian Postage Stamp (2003) commemorates the millions of victims in the famine of 1932–33. © Ukrainian Postal Service (public domain).

Kharkiv was at the center of the whirlwinds of the purges and Great Terror, with UFTI suffering arrests and executions. The ferocity of the purges may have to do with the desire to end Ukrainian nationalism. Many Ukrainian universities were transformed into pedagogical institutes with emphasis of vocational and Marxist instruction. Party and university officials expelled faculty members suspected of Ukrainian nationalist sympathies. They determined next to clean UFTI of alleged enemies who were identified as members of fabricated conspiracies, were foreigners or had foreign contacts, or faced a litany of other charges. In October 1935 they purged the institute of so-called hostile class and counterrevolutionary elements.[3] Future Nobel laureate Lev Landau fled to Moscow to be arrested there, Shubnikov and others were arrested and tortured. Landau and Shubnikov had made it difficult for themselves. They let it be known that they disliked Kharkiv. They considered it to be a provincial outpost of industry and agriculture, not of science. The university administration fired Landau for his suspect political views, his haughty disregard for students who did not measure up to his standards, and his haphazard attitude toward work. In solidarity, Landau's university colleagues sent protest letters, an act that was taken to be an illegal strike.[4] Such loyal communist scientists as Leipunsky faced the constant threat of arrest. At interrogation under duress he admitted inadequate vigilance toward ever-present dangers of foreign specialists at UFTI.[5] By 1937–38, UFTI fell directly into the Stalinist maelstrom. NKVD officials identified a nest of anti-Soviet activity at UFTI. They singled out a series of party members, Jews, "careerists," foreigners, and members of the Trotskyite opposition, and counterrevolutionaries. Dozens were purged, and the institute was decimated.[6]

Thus, the rise of nuclear physics and technology occurred against the backdrop of political turmoil, cultural revolution, famine, and the Great Terror, demonstrating that the intellectual and political leaders of the USSR were able to support and expand scientific activities even in a time of upheaval. Cities swelled in population overnight, while violence racked the countryside, largely because of the Bolsheviks' insistence to push peasants into collective farms, confiscate their grain, and summarily force ten million people to go into exile in the Far

North, Siberia, or the Far East. Millions died of disease, dehydration, and starvation en route to labor camps and work sites in overpacked trains, and perhaps three million Ukrainian peasants died within the republic's borders of famine. The physics community, too, suffered directly and indirectly, with purges hitting such fields as astronomy and such specializations as theoretical physics with catastrophic losses, and autarky was imposed on the scientific community.[7] Yet because of the increasing support to the sciences for the industrialization effort, and perhaps because they could find solace in their work to avoid dwelling on the purges and the loss of family and friends, specialists continued to undertake cutting-edge research in a variety of fields: low temperature physics, theoretical physics, solid state physics and nuclear physics among them, until the Second World War disrupted their efforts.

The Second World War and Ukrainian Physics

The Second World War dealt another major blow to Ukrainian physics and the burgeoning nuclear enterprise. When the Nazis invaded, Kharkiv physicists evacuated what they could just ahead of the onslaught. They were able to disassemble some of the equipment, and cart it off to Almaty, Kazakhstan, where they set up quarters in a few rooms of the university. Here, after the war and on the orders of Igor Kurchatov, they contributed to the initial steps of the atomic bomb project. They studied the theory of moderation and scattering of neutrons in crystals, especially in graphite, with Alexander Akhiezer and Isaak Pomeranchuk, two leading UFTI theoreticians, completing the first Soviet monograph on nuclear theory.

But Kharkiv was captured by German armies on October 24, 1941, and a disastrous Red Army offensive failed to retake the city in May 1942, with at least 300,000 soldiers captured, taken as POWs, executed or killed in battle.[8] Again the Red Army retook the city in February 1943, saw it retaken in March 1943, and finally retook it late in August 1943. Seventy percent of the city was destroyed, the population was halved, and a vital Jewish community was liquidated by the Nazis.[9] Indeed,

Stalin was unprepared for the Second World War, had negotiated a non-aggression pact with Hitler in 1939 that enabled the two dictators to divide Central Europe with their armies, and he seemed incapable of leadership in the first weeks after the Nazi invasion of June 22, 1941. As a result, twenty-five million Soviet citizens died in the war.

After the war, the physicists returned to find their city and institute in rubble. The institute's main building had been blown up by the retreating Nazis. All the windows were broken, the contents ransacked. There was no heating fuel. The plumbing and sewer systems were destroyed. Sinelnikov found his flat absolutely empty and filthy, his beautiful Steinway piano on its side having been used by the Germans as a platform for washing lorries. Luckily, Sinelnikov and Valter reported to Kurchatov that the electrostatic generator at UFTI was only slightly damaged and could be repaired. The vacuum system seemed to be in full order, with electrical equipment of the generator practically in working condition. Sinelnikov asked Kurchatov to authorize funding and supplies needed to reestablish a fully operable workshop to repair or replace this equipment and to reestablish normal supply of gas. Yet problems in securing the early release of several physicists from service in the Red Army and shortages of supplies including special rations delayed the completion of repairs for some months.[10]

Most observers assume that the centralized nature of the Stalinist system enabled economic planners to allocate resources in a rational fashion. But the physicists at UFTI knew firsthand that Party officials played favorites in rationing personnel and equipment among the growing number of research institutes of the scientific establishment after the war, especially in the face of endemic postwar shortages from building supplies, to steel, concrete, and food. Even UFTI's position as a central institute of the atomic bomb project did not guarantee supplies. Having returned to Kharkiv after the Nazi withdrawal, Anton Valter and his associates set to work, getting some of the equipment back on line by 1946, including two electrostatic generators that were, to his understanding, the only such functioning devices in the USSR. Yet funds were insufficient to expand crucial research. To make matters worse, of the twenty-nine specialists recently graduated from

UFTI programs on nuclear themes, not one gained employment in the institute; they were all transferred to Moscow. As late as January 1951, Anton Valter addressed Lavrenty Beria about the unsatisfactory level of support for his sector that delayed R and D on electrostatic generators. He asserted that UFTI should be entitled to keep some of its own equipment for research programs, not serve as a production bureau for other institutes of the nuclear enterprise that needed generators. Valter put it simply: "From my point of view it would be significantly undesirable to transform the role of [my] department … into the role of a design production outfit for … electrostatic generators for other organizations."[11] In February 1952, Kirill Sinelnikov, now director of the UFTI, again wrote Beria to protest the difficult financial straits of his institute. Sinelnikov was a central figure in the Soviet nuclear program. Sinelnikov had the distinct impression that the physicists of UFTI (called Laboratory No. 1 of the atomic project) had been slighted by Beria. Accordingly, their efforts to expand research in high energy and nuclear physics, including a new program in fusion (controlled thermonuclear synthesis) lagged. Funds indicated for building apartments had not been spent, making it impossible to attract younger specialists. Space in the city rebuilding from the war's destruction remained at a premium.[12]

As if the challenges of recovering from the war were not bad enough, the authorities chose the postwar years as the time to renew the call for ideological vigilance in science. In every scientific discipline, in every region of the country, Marxist philosophers and scientists of firm Stalinist conviction asserted their authority in science. In physics, this meant interest in how relativity theory and quantum mechanics were commensurate with the Soviet philosophy of science, dialectical materialism. Sinelnikov assisted Akhiezer and Valter in developing readings in the Marxist philosophy of science for young students preparing for qualifying examinations for the candidate of science degree. In biology, Trofim Lysenko, who rejected genetics, gained the authority to ban genetics from textbooks and banish researchers from their jobs at a national conference in 1948. Such a conference was planned in physics as well for some time in 1949 to condemn relativity and quantum mechanics. But by the grace of Kurchatov and the atomic bomb project the enterprise saved from an outcome similar to that in genetics,[13] although many physicists' careers were damaged.

Still, nuclear physics and nuclear engineering grew rapidly in Ukraine in the postwar years—as they did in other union republics. As part of "Atoms for Peace" (see Chapter 3), the nuclear establishment worked with national and republican officials and scientists to open new research institutes and training programs in universities. From central design and construction institutes, they disseminated research reactors, particle accelerators and other equipment crucial to expansion of the nuclear enterprise. From the 1950s and 1960s the Soviets built and sold IRT-2000 reactor (generally 2.0 to 2.5 MW) to Moscow, Tomsk, Sverdlovsk, Salaspils, Latvia, Tbilisi, Georgia, and Dubna facilities. They commissioned the VVR-M reactors in a small (up to 0.5 MW) and scaled up (10–15 MW) design in Gatchina, Obninsk, Kyiv, Almaty, Kazakhstan, and several East European socialist nations. But the first efforts in nuclear engineering were focused on the atomic bomb, especially after the US dropped bombs on Hiroshima and Nagasaki, Japan, in August 1945.

CHAPTER 2
NUCLEAR DEFENSE

In September 1949 western newspapers screamed the headlines of the first Soviet atomic bomb; spy planes had detected the telltale radioactivity in the atmosphere. This startling news worried many people in the democratic world who already saw things in black and white: evil, godless communism versus just and free democracy and capitalism. The West no longer monopolized a weapon they believed would curb Soviet aggression. On September 23, 1949, President Harry Truman tried to calm concerns about the Soviet achievement, first contending that western specialists had anticipated this even for several years, and then reiterating a call for effective international control of the weapon.[1]

And, indeed, Truman was correct that the Soviet bomb should not have been a surprise. The Soviet Union had worked feverishly on developing it since August 1945 when the United States dropped uranium and plutonium bombs on Hiroshima and Nagasaki. Stalin and his advisors would not tolerate a US monopoly on nuclear weapons. Marshalling all available resources—manpower directed to work in a command, centrally-planned economy capable of allocating scarce resources, prisoners of war and gulag inmates into ever-growing constructions brigades, and drawing on the pre-war nuclear establishment, scientists gained carte blanche to chart uranium reserves, acquire equipment, expand institutes and programs, and requisition talented young researchers. The bomb project was guaranteed success. The Soviet atomic bomb project, from start to finish, took roughly the same time as the US Manhattan project, from drafting of personnel to the equipping of dedicated laboratories 1 and 2, the latter now the Kurchatov Institute for Atomic Energy (KIAE), and from the bringing on line the first atomic "pile" (reactor) within

Moscow city limits to the detonation of the first bomb in a Kazakh desert in August 1949. While some Cold Warriors argue that the rapid Soviet success was largely the result of espionage, it is clear that a superior foundation of physics research in Kharkiv, Leningrad, and elsewhere, and the command system of economic supply, including gulag-supplied uranium, construction and other components of the bomb project, were keys to the entirely indigenous effort. At the end of the war, even Nazi materiel, uranium, and scientists were spirited out of the Reich for use in nuclear facilities in the USSR. It helped that Stalin and Beria ordered that no obstacles of labor, safety, and supply stand in the way of the project.[2]

Over the next fifteen years the nuclear weapons enterprise expanded rapidly in a series of newly established, closed cities dedicated to the bomb (the "Closed Administrative-Territorial Formation," or ZATO in its Russian acronym). The citizens of these cities, often designated by a post office box number, lived a life of Soviet luxury; yes, their cities were closed to outsiders and they themselves could not move away without permission, and they needed special passes simply to leave and return on any business. But their schools and hospitals were better than the norm, and their stores were well stocked. The residents bought fully into the need to build more and better bombs, not only because of the quality of life, but because they firmly believed they were defending the motherland against the ever-present American threat. In his *Memoirs* Andrei Sakharov, the father of the Soviet hydrogen bomb project and later dissident, discussed his experience in the "installation" in Arzamas, his acceptance of the daily routine where he worked in an institute surrounded by barbed wire, where prisoners were marched in columns—in rags—to engage in construction. While the scientists had a free hand in their own work, there was a standard fare of borders, boundaries, gates, passes, and prisoners in ZATOs.[3] A friend who lived in a ZATO told me that he had a pump-up rubber dingy in his eighth-floor apartment. He awaited an American nuclear blast in the harbor, and when the radioactive tidal wave came rushing ashore toward his building, he would launch himself and his family out the window to float off to safety. ZATOs, strictly speaking, are not a postwar phenomenon, but date to the first-five year plan and the re-introduction of the Tsarist system of internal passports, and their appearance was tied to the Soviet practices of ubiquitous ID cards and labor books.

Letter to Stalin

In 1989 in Leningrad (Saint Petersburg), I met Georgii Flerov to discuss his work in nuclear physics. An academic, Lenin Prize laureate, and director of a laboratory at the Joint Institute of Nuclear Research in Dubna where he worked on synthesizing and studying transuranic elements, Flerov discovered spontaneous fission in 1940 with Konstantin Petrzhak, a laboratory head at the Radium Institute in Leningrad, with a device that they placed 60 meters underground in the Moscow Dinamo Metro station. But Flerov wanted to talk with me about two letters he wrote to Stalin in 1942, urging him to start work on a "superweapon," an atomic bomb, and gave me copies of the letter, now widely published.[4] Flerov's letters reminded me of that penned by Leo Szilard and signed by Albert Einstein to President Franklin Delano Roosevelt of August 2, 1939, that urged the president to commence a crash atomic bomb project in the United States over fears that the Nazis were surely working on such a terrible weapon and would use it during war.

Flerov worried about military possibilities of fission because German, American, and British scientists had ceased publishing on the subject. Were they developing an atomic weapon? He wrote Kurchatov first about his worries but, getting no answer from the busy Kurchatov, he wrote Stalin in 1942—who still took little notice. On top of this, the Soviet program was slow to begin because of the Nazi invasion, the hasty evacuation of Leningrad institutes connected with nuclear research to Kazan, and the falling of Kharkiv and Kyiv into Nazi hands. Indeed, Kurchatov was involved with anti-mine research and had moved from Kazan to Murmansk for work with the Northern Navy. The Soviets were aware through espionage by late in the war, however, of the US Manhattan Project. Government officials agreed to set up a uranium commission and to study the possibilities of isotope

Figure 2 Physicist Georgii Flerov wrote Joseph Stalin in 1942 urging a crash nuclear bomb project. This Russian postage stamp (2013) honors the anniversary of Flerov's 100th Birthday. © Russian Postal Service (public domain).

separation and a chain reaction, and at the end of 1943 they appointed Kurchatov to investigate a bomb. When he learned of the bombing of Hiroshima and Nagasaki, Stalin ordered acceleration of Soviet efforts, assisted by German scientists who had been captured at the end of the war and through continued intelligence efforts.[5]

Reminiscent of the Allies' "Project Paperclip" and "Alsos" operations which were advance parties of soldiers with scientists sent into the crumbling Third Reich to apprehend Nazi researchers to employ in the western space and bomb programs (e.g., Werner von Braun, a leading architect of NASA's "Mercury" program, was a Nazi co-designer of the V-2 World War II rocket), so Soviet scientists, soldiers, and their mission seized Nazi nuclear research facilities, entire libraries, and scientists to employ in their bomb project.[6] They simultaneously gained control of the Ertzgebirge uranium mines which they exploited to the fullest, exporting ore to the USSR.

The Soviet program involved scientific research and testing in a series of newly-founded laboratories; accelerated efforts to find uranium deposits; the building of experimental and then production reactors to understand fissile material better and produce plutonium; and massive ore processing and enrichment factories. These efforts required vast land takings including three major testing grounds in the Ural Mountains, Kazakhstan and the Arctic archipelago of Novaia Zemlia. The secret police and military managers of the project were ruthless in pursuit of the bomb; the human costs in terms of prisoners and prisoner deaths, employees and scientists in ZATOs, as well as downwinders and soldiers, exposed to radiation, and the level of environmental degradation are impossible to determine.

Even before Flerov's letter, in June 1940, the leadership of the Academy of Sciences appointed a Uranium Committee with Ioffe, Petr Kapitza, Aleksandr Fersman, Kurchatov, Yuli Khariton, and others as members. Moving slowly, some two years later the State Defense Committee began to focus on study of nuclear energy, and in February 1943, it determined to build a nuclear weapon, ultimately under the direction Beria, and with Kurchatov the scientific director of the project.[7] The major institutional homes of the Soviet bomb effort were Laboratory 1 (UFTI) and Laboratory 2 (Kurchatov's new institute, later KIAE). At first, a series of tents in what was then a large open field 12 kilometers (km) to the northwest of the Kremlin ("October Field"), KIAE is now deep within Moscow city limits, surrounded by apartment buildings and businesses, and at one time had twelve operating reactors, until recently six, although at present none is operational. Moving ahead rapidly in December 1946 Kurchatov started up the first research reactor (the "F-l"), a simple graphite hemisphere. In the 1990s I had the honor of being escorted by Kurchatov's deputy, Igor Golovin, to the F-1, still operational and warm to the touch. From the time the F-1 first produced a chain reaction to the time of the first Soviet atomic bomb test, "First Lightning," in Kazakhstan in August 1949, almost exactly the same number of days passed as in the United States from the time of its first pile in a squash court at the University of Chicago to the explosion of Trinity, the first US test, at Alamagordo, New Mexico, in July 1945.[8]

Figure 3 Igor Kurchatov, head of the Soviet Atomic Bomb Project, later pushed a bold program of nuclear power. This Russian postage Stamp (2003) honors his contributions. © Russian Postal Service (public domain).

On paper, Laboratory 2 was created in September 1942 by a resolution of the State Defense Committee (GKO) "On the Organization of Work on Uranium" that put Ioffe's LFTI, now in evacuation in Kazan, in charge of a special atomic laboratory with ten men working under Kurchatov. In February 1943 it was decided to create Laboratory 2 on the basis of this group to produce nuclear weapons. Research on nuclear reactions commenced in Kazan from 1943, then in a branch of Laboratory 2 in Leningrad and another in Sverdlovsk. Eventually, all work was concentrated in October Field, close to Stalin's scrutiny. Buildings for the laboratory and housing for workers were built rapidly in the postwar years, secured by the ruthless Beria.[9]

Beria, Stalin's longest-serving secret police chief, was by all accounts an able administrator with a tremendous memory and attention to detail, and after Stalin's death he pushed to empty the prison camps even before the reformer, Stalin's successor, Nikita Khrushchev, determined to do so.[10] Beria was also a murderer who personally signed death lists, ordered the slaughter Polish officers at Katyn during the Second World War, willingly carried out purges of entire groups of citizens, happily oversaw the expansion of the gulag system, and he was a rapist. As for the scientific sphere, Beria organized special camps, called *sharashki*, to gather scientists together in research programs, especially for the Soviet aeronautical program.[11] Arctic and Siberian camps frequently involved expeditionary work on top of cruel road, railroad and mine construction; those camps exploited geologist and geophysicist prisoners. The major *sharashka* for nuclear specialists was Laboratory B in Sungul, a resort in the Ural Mountains, that operated from 1946 until 1955, when it was disbanded, and some of its staffers were employed in NII-1011 (the All-Russian Scientific Research Institute of Technical Physics, VNIITF, created as a backup to Arzamas) and others were transferred to Combine 817 (Maiak in Ozersk).

At its founding, "B" had two departments: biophysical, headed by the renowned N. V. Timofeev-Resovsky, and whose work embraced radiation genetics, population genetics and microevolution, and radiochemical, headed by S. A. Voznesensky. Timofeev-Resovsky worked as director of the Genetics Division at the Kaiser Wilhelm Institute for Brain Research from the early 1930s until 1945—even with Nazi funding, and he was repatriated and imprisoned when the Soviet scientific search team arrived in Berlin in May 1945, with nuclear physicists Iulii Khariton, Isaak Kikoin, Lev Artsimovich, and Flerov identifying scientist-prisoners. (Khariton located 100 tons of uranium oxide that was secreted back to Moscow.)

Recognized for his important radiobiological work, Timofeev-Resovsky was sent to laboratory B.[12] The team also found German nuclear specialist Nicholas Riehl who served as B's scientific director; according to some sources Timofeev-Resovsky apparently persuaded a number of German nuclear specialists to go with the Soviets, not to the Americans in their mission. But first Timofeev-Resovsky was sent to the gulag and when released to return for scientific research he was in such a bad state that he could not remember his sons' names. Before 1955, from 200 to 500 employees worked in "B" that included imprisoned Soviet scientists and German internees.[13]

In the early 1950s a struggle against so-called cosmopolitanism (having feelings of belonging to the world community, not dedication strictly to Soviet causes, for example, Jews who were immediately suspect of Zionism and allegiance to Israel) intensified with the uncovering of the "Doctors' Plot." According to Stalin and his paranoid collaborators, Jewish physicians intended to poison the Kremlin leadership. Dozens of leading physicians disappeared. Only Stalin's death prevented another horrible purge that, apparently, would have extended far into the population.[14] In this atmosphere a colleague of Sakharov's at the installation, Mattes Agrest, lost his job, but only his job. In his ID papers he indicated that at the age of fifteen in 1930 he graduated from Hebrew seminary and received a rabbi's certification. The KGB agents were horrified to discover their oversight, and, even though Agrest had given decades of his life to the state, they banished him immediately from Arzamas. Igor Tamm and Sakharov intervened to get Agrest a few days to gather his things. The Doctors' Plot led to the removal from Arzamas of many other senior scientists. Throughout the country a number of Jewish specialists were fired from the atomic bomb project.

In 1945, to create another site both to have competing research teams and calculations and to ensure program continuity should Laboratory 2 fall under attack, the authorities established Arzamas-16 (originally KB-11, now the All-Russian Scientific Research Institute of Experimental Physics [VNIIEF]) in a former monastery in the city of Sarov under the direction of Iulii Khariton. Other leading specialists who would work at Arzamas were Igor Tamm, Andrei Sakharov, Iakov Zel'dovich, and Isaak Pomeranchuk. Khariton's father was arrested by the NKVD in 1940 in Latvia after the Red Army invaded and died in the gulag, while Khariton remained faithful to weapons development for almost fifty years at Arzamas as its director. At first the scientists at Arzamas and in Moscow doubted they might build a bomb with weapons-grade uranium and focused on plutonium, which was, in fact the fuel for First Lightning—the first Soviet bomb. The Hiroshima and Nagasaki attacks indicated both were feasible, but KB-11was set up to work on a plutonium weapon.

There was no respite for scientists. If Stalin had ordered scientists be recompensed well for their labors on behalf of the state, they still faced constant verification of their ideological soundness and their loyalty to the state. Cold War xenophobia led to the so-called *Zhdanovshchina* in the late 1940s, an attack on western influences, on "kow-towing" before the west, or "cosmopolitanism." No one was immune. Such writers as the satirist Mikhail Zoshchenko, such composers as Dmitrii Shostakovich and Sergei Prokofiev, geneticists, and theoretical physicists, hundreds of elite and ordinary citizens again disappeared into the maelstrom of Stalinist politics.

Around this time, Moscow University and other physicists mired in mechanistic understandings of the world, sought a national conference to condemn the philosophical failings of relativity theory and quantum mechanics. This conference would have the same result that a 1948 conference did for biology, proclaiming the quack biologist Trofim Lysenko and his anti-Darwinian theories victorious

Figure 4 Iulii Khariton was the chief Soviet nuclear weapons designer for fifty years. This commemorative Postage Stamp (Russia, 2004) celebrates the anniversary of his 100th birthday. © Public domain.

over genetics. According to an apocryphal story, Igor Kurchatov and his colleagues gathered to discuss how to prevent the condemnation of relativity theory and quantum mechanics. They called Beria to inform him that condemnation would mean no bomb could be built. Beria relayed this information to Stalin who said, "Tell them to build the bomb. We won't hold any conference. Besides, we can shoot them all later."

Administrators and Spies

For the US Manhattan Project, J. Robert Oppenheimer was the scientific director. General Leslie Groves of the US Army Corps of Engineers was the chief government military administrator. They had a significant degree of respect for each other, if at times a challenging relationship. But Oppenheimer capably directed the project to the detonation of the first US bomb in the Trinity Test on July 16, 1945, with bombs dropped on Hiroshima and Nagasaki in early August, and Groves saw to the development of major national laboratories and factories to produce weapons grade plutonium and uranium, with the scientific center of the project Los Alamos National Laboratory (New Mexico), Oak Ridge (for uranium enrichment, Tennessee), and Hanford (for plutonium production, Washington). Kurchatov was Oppenheimer's counterpart, and Beria was Groves's.

Beria, with his crushing orders and access to prison labor, managed the project well. He had the requisite skills for managing the project: energy, clear thinking, administrative purposefulness, excellent subordinates, plus access to hordes of prisoners. In addition, Beria used the NKVD world spy network to infiltrate western (German, American) projects. The USSR had a toehold in the United States in the Amtorg Trading Corporation that engaged in industrial espionage, and its representatives coopted American sympathizers into the Soviet spy effort. There was a spy in the Manhattan project: Klaus Fuchs, a German theoretical physicist who contributed to solving the problem of plutonium implosion. He supplied information from the American, British, and Canadian Manhattan Project to the USSR from his post at Los Alamos National Laboratory, then returned to the Great Britain to work at its Harwell research center. Discovered, arrested, and tried by the British, he was convicted of espionage in 1950, served nine years in prison, and then was released to East Germany where he resumed his career, dying in 1988. Most analysts agree that Fuchs saved the Soviets time, but had not given them the bomb on a plutonium platter which they would have succeeded in building in any event. Surely though Fuchs assisted in speeding the Soviets early to a plutonium device, and he helped in determining critical mass, with a plutonium production reactor an easier path to weapons grade fuel than uranium isotope separation and enrichment.

Another way in which the Soviets were fully aware of the US bomb program was the fact that western publications included the so-called Smyth report make available in Russian by the end of 1946.[15] Sakharov was shocked to learn about Hiroshima and Nagasaki in 1945. When he finally got to read Henry Smyth's *Atomic Energy for Military Purposes* (1945) he made preliminary calculations about the potential for a bomb. He and hundreds of other promising young scientists were drafted into the atomic bomb effort. The secrecy regime introduced at most institutes was immediate. The easy-going culture gave way to strict control over all information. Scientists had to number every page of their notes which were collected at the end of every day to be stored in safes. The custodial crews washed the blackboards every night. The KGB increased its presence. Los Alamos was open in comparison.

From First Lightning to Serial Production

On August 29, 1949, the Soviet Union secretly conducted its nuclear test, "First Lightning" at the Semipalatinsk Test Site in Kazakhstan. Legend has it that the scientists were fearful of failure because this would mean certain retribution from Stalin and Beria. But in mid-August 1949 Kurchatov gathered his team gathered nervously for the first Soviet bomb (RDS-1). The Soviets shipped the bomb components 3,000 kilometers from Arzamas to the Semipalatinsk, Kazakhstan, test site. The bomb was detonated from a tower. Nearby, in addition to instruments, the scientists built wooden and brick houses, bridges, tunnels, water towers, and even a simulated subway station to learn about blast, heat and radiation impacts. Not for the last time, they distributed over 1,500 caged animals in the predicted bomb zone. Kurchatov arrived in May with other scientists and officials for a series of dress rehearsals. Beria showed up in mid-August and frequently reported to Stalin by telephone. The 22 kT device, RDS-1, with a solid plutonium core, was an implosion device similar to the Fat Man dropped on Nagasaki. At 6 am the bomb was detonated as Kurchatov, Beria and others watched from the command post. The fireball and mushroom cloud, the blast wave and the debris brought euphoria to the team. Khariton recalled that Beria hugged Kurchatov. Khariton later said, "When we succeeded in solving this problem, we felt relief, even happiness—for in possessing such a weapon we had removed the possibility of its being used against the USSR with impunity."[16] Five RDS-1 weapons were completed as a pilot series by March 1950 with a serial production of the weapon that began in December 1951.

Two years later, in autumn 1951, the Soviets detonated the 38.3 kiloton RDS 2 ("Joe 2" by the CIA), a tritium boosted uranium implosion device and then the RDS-3 with a uranium shell and plutonium core that was the first Soviet air-dropped bomb test. Released at an altitude of 10 km, it detonated 400 meters above the ground. RDS-4 was, fatefully, the result of research on small tactical weapons—that might be used in a field battle, fatefully because in 1954 a similar bomb was dropped by a Tu-4 bomber during the Snowball exercise at Totskoe among 40,000 infantry, tanks, and jet fighters. Many of these nuclear veterans suffered health consequences and early death; the Americans and French also tested nuclear bombs on their own soldiers as guinea pigs. With the RDS-6 the Soviets tested a hydrogen bomb ("Joe 4") of "layer-cake design of fission and fusion fuels (uranium 235 and lithium-6 deuteride) [that] produced a yield of 400 kilotons."[17]

The USSR's scientists and military leaders worked feverishly to design and test nuclear bombs with larger yields, more or less radioactivity, in the atmosphere, on the earth's surface and underground, and even for peaceful purposes of excavation, creation of underground domes for storage, building dams and canals, and putting out oil well fires, for a total of 981 devices, a total yield of 300,000 kTs, and already with 135 of them in 1961–62, or one-seventh of the total, ready to be used at a time of great international tension over the Berlin Wall and the Cuban Missile Crisis, and when the world's citizens, including those in the USSR, anticipated nuclear annihilation.[18]

Weapons grade fuel came from a variety of facilities and factories in ZATOs—Krasnoiarsk-26, Tomsk-7, Cheliabinsk-40. Like such massive uranium and plutonium production facilities in the United States as Oak Ridge, Tennessee, and Hanford, Washington, Sellafield in the United Kingdom, those in the USSR stretched to the horizon. For a sense of the scale and challenges in operating such a facility, consider the Maiak Chemical Combine in Cheliabinsk. Engineers quickly built a series of massive, but environmentally unsound plutonium production reactors, uranium isotope separation and enrichment facilities, and fuel fabrication plants. Maiak, in Cheliabinsk for manufacture of plutonium, was built by 40,000 gulag prisoners and POWs. The plant occupied 90 square kilometers. Haphazardly managed high- and low-level radioactive waste dumps serving Maiak and dozens of other facilities filled the Urals region. The decision to locate these facilities in the central Urals region near metallurgical, construction, and chemical factories of Cheliabinsk, Perm, and Sverdlovsk provinces was based on strategic considerations—they were far from the USSR's borders, and proximity to industry and employees—industry expanded rapidly during the war. Another facility, the Ural Electrochemical Combine, established in 1946, used gaseous diffusion to separate uranium isotopes.[19] Its main building stretches one km and houses thousands of industrial centrifuges. A sixth generation of serial centrifuges was completed in the early 1980s.[20] As with chemical fertilizers, asbestos, steel, and coal, and other products produced in the Urals region, rapid production, and large scale were central to design of nuclear facilities, and safety a second thought.

By the late 1950s production of warheads moved into the industrial mode with plutonium and other production plants pumping out highly radioactive waste in addition to weapons grade fuel. Citizens and landscapes paid the price. The settings for bomb production—mines, uranium ore processing mills, plutonium production reactors, fuel fabrication facilities, and so on—distributed their waste into nearby rivers and lakes, or buried it haphazardly underground, or stored it in fragile, single-hull tanks doomed to failure. Through a series of spectacular and not-so-spectacular accidents, and through leaks and gravity, the carcinogenic and mutagenic radionuclides irradiated tens of thousands of citizens, some of them willing participants in the arms race who discounted or ignored the dangers, and many others unwilling and ignorant of the daily risks of living in Nuclear Russia. To this day, Russia is struggling with the legacy of radioactive waste from the Arctic region to the Urals, from St. Petersburg and Moscow to Siberia, and citizens have largely been denied participation in managing this legacy.

As in the United States, indigenes were removed from their homelands to facilitate the Cold War arms race, Bikinians for the United States, Nenets from Novaia Zemlia and Kazakhs from Semipalatinsk, and "downwinders" exposed to radioactive fallout in both nations. Red Army soldiers were ordered into ground zero minutes after tests so that the generals could plot the use of nuclear weapons in war. And, as in the United States, the legacy of nuclear waste and fallout from the production facilities, tests and PNEs continues to plague the Russian nation, even as Rosatom has embarked on massive programs to expand the nuclear enterprise in the 2020s. What did citizens know about the arms race? What was it like to live in a nuclear polygon? What did they think of their nuclear heroes in the Cold War? What role did these scientists play in the Gorbachev period of reforms and in the wild nuclear west of the 1990s when many people worried that weapons and weapons expertise might migrate to "rogue" nations or terrorists? And what is the place of nuclear memory among indigenes today?

Nuclear Guinea Pigs

The Soviets built three weapons test sites: at Semipalatinsk (18,000 square kilometers) in Kazakhstan where 456 tests were carried out (340 underground and 116 atmospheric tests suspended from towers or dropped from aircraft), Novaia Zemlia in the Arctic Ocean with 130 tests, and a site in the Ural Mountains. The vast expanse of lands taken by the state for weapons testing indicates the importance of nuclear weapons for Soviet power. Semipalatinsk had four major testing areas at the site, along with two research reactors, supported by the ZATO "Kurchatov," at one time a city of 20,000, and now a decaying monument to Soviet WMD. After the Limited Test Ban Treaty (1963), the Soviet Union carried out 340 more underground nuclear tests in caves or boreholes at all four sites. Semipalatinsk also was the location of nine of the Soviet Union's peaceful nuclear explosions (see below). The scale of the Degelen Mountain nuclear test facility at Semipalatinsk, the largest underground nuclear test site in the world, consisting of 181 separate tunnels, gives a sense of the resources the USSR devoted to nuclear testing. From 1997 to 2000, these tunnels and associated shafts were sealed as part of a joint US-Kazakhstan program under the Nunn-Lugar Cooperative Threat Reduction program.[21]

The absence of proper safety measures at the Semipalatinsk test site became clear when the Soviet Union collapsed in 1991 and local activists, NGOs, and western observers began to agitate for remediation of the radioactive and other toxic wastes from forty years of secret bomb testing activities. First, there was no perimeter fence around the site that prevented uninformed local people to enter the site. Security forces were assigned to protect the test reactors only in the 1990s. Additionally, beryllium, coal, and gold were mined throughout the site and table salt was produced from a lake located near the main test field.[22]

Testing had significant environmental and public health impacts, at first because of atmospheric testing, later because of the venting of radioactive gases until the last tests in 1989.[23] Like French military authorities for their Tahiti tests and US officials for their Bikini Atoll explosions, so Soviet officials ignored local people, describing the Kazakh steppe surrounding the Semipalatinsk site as "uninhabited" while small villages were on the site's outskirts, and Semipalatinsk (today, Semei), a major stop on Turksib Railroad and at its maximum with 300,000 inhabitants, was only 160 km away. Villagers who lived close enough to witness the tests were never informed about the nature of the tests or the dangers they faced. The total number of people exposed to "substantial radiation doses" from 1949 to 1962 is estimated by Kazakhstan's Institute of Radiation Medicine and Ecology to have been between 500,000 and one million. The extensive contamination resulted in high incidences of cancer and mortality among nearby Kazakh residents.[24] Many of the long-lived radioactive isotopes will persist for centuries and the local villages have abnormally high rates of cancer, especially breast and pulmonary, and birth defects.

Another important moment in Nuclear Defense was a test conducted at the Totsk site in the Arinbuk region of the Southern Urals on September 14, 1954, that exposed 44,000 soldiers to excessive radiation. The test was like several of those involving "nuclear guinea pig" soldiers in the United States, in Algeria by the French, and in Australia by the British. (Some 400,000 US soldiers were exposed to some of the 200 US atmospheric conducted from 1946 to 1962 by the US either as observers or in cleanup.[25]) At Totsk the Soviets tested relatively small nuclear weapons that might be used on the battlefield to destroy enemy soldiers, yet would run the great risk of exposing one's own soldiers to blast, heat, and radiation. Even people living in the blast test region remained exposed to ionizing radiation forty-two years later including plutonium and [137]cesium. The authorities required those involved in the test to sign secrecy agreements. The soldiers were dressed in ordinary uniforms, rubber boots and gas masks, without any shelter, to watch a 40 kt bomb explode above their heads, about 3 km away. Then the soldiers engaged in war games to see how men and equipment faired under "nuclear attack." They entered a region of Potemkin houses, fortifications, and military vehicles. Reporters of a Russian investigative television program, "*Kak Eto Bylo*" (As It Happened), revealed that Soviet military officials had exposed soldiers and pigs to the same kind of experiments that US military officials had in the 1950s. Years later many of them died from cancers and other illnesses.[26]

A variety of farm and other animals were also anchored in the test zone to subject them to blast, burn, and radiation and evaluate the impact of nuclear weapons on them. The Americans carried out similar tests, for example, in Operation Plumbbob, over a nearly five-month period in 1957 at the Nevada test site that released vast amounts of radioactive iodine and involved over 1,200 pigs subjected to biomedical experiments, for example, dressing them in clothing made of different materials to test blast resistance and blast-effects.[27]

Tsar Bomba and the Novaia Zemlia Arctic Archipelago

Under Nikita Khrushchev, the Soviets unilaterally imposed a test ban for over two years from 1959 until 1961. Perhaps because of the Francis Powers U-2 spy plane incident when Powers was shot down over the USSR, and also because of the Berlin and Cuban Missile Crises, not to mention the Bay of Pigs fiasco, the Soviets determined to test a superbomb of unsurpassed power. If in autumn 1959 Khrushchev had had a successful US visit and had built good relations with President Dwight Eisenhower, then by the time of the Powers incident in May 1960 future meetings, including a visit of Khrushchev to Moscow, were cancelled. On top of this, the United States used the moratorium to increase the number and total megatonnage of its nuclear arsenal, while the USSR had mastered thermonuclear charges but serial production lagged as did weapons-grade plutonium production capacity. Maiak was in operation, but facilities at Tomsk and Krasnoyarsk were as yet to operate at full power. Soviet leaders decided to respond "asymmetrically by developing superpower thermonuclear weapons in order to parry the considerable superiority of the U.S. thermonuclear arsenal." Khrushchev announced this decision "to leading Soviet nuclear physicists at a closed-door meeting in the Kremlin on July 10, 1961."[28]

For Tsar Bomba the goal was to intimidate the enemy. For such a huge weapon, planners chose the remote Novaia Zemlia archipelago with an area of 85,000 square kilometers. Considering Nenets inhabitants unimportant, they ordered the complete closure of the archipelago to indigenous life. The authorities moved 298 people away, some to Arkhangelsk, some to a god-forsaken military airbase at Amderma, a town near Vaigach Island founded by gulag prisoners in 1933, and others to Kolguev Island where they might continue their reindeer herding activities.[29] The Nentsy were promised jobs and monthly payments, but in fact received little support from the state. The construction of the Chernaia Guba site in the archipelago began in summer 1954, with ten construction battalions assigned to it. The first study of underwater nuclear explosion was planned for September of the following year. Workers lived in tents through winter to prepare the site; eventually they built residential buildings, power, and water supply services, communications, scientific facilities, material supply services, and runways and docks. Fighter aviation regiments and ships were assigned to defend secret facility.

A total of three sites were deployed on Novaia Zemlya: Sukhoi Nos, Chernaia Guba, and Matochkin Shar. Chernaia Guba was active between 1955 and 1962 to carry out low- and medium-yield atmospheric explosions with underwater and surface nuclear tests. From 1964 through 1990, nuclear tests were conducted in deep underground shafts on the Gulf of Matochkin Shar's southern bank. Between 1957 and 1962 until the signing of LTBT, atmospheric tests were conducted on Sukhoy Nos Peninsula, north of Matochkin Shar Strait. To handle all this activity, the Belushia Guba settlement, located on the archipelago's southern island, the administrative center and the headquarters for the military base, housed over 2,000 people, nearly all of them military personnel and their families who were strictly devoted to the atom.[30]

Some of the tests, of course, resembled those of the United States because of the simple need to understand how to manufacture devices that met the needs of burst, radiation, and shockwave to destroy the enemy, his ability to strike back, and the potential impact on entire civilian populations. The scientific make-up of tests only partially shielded their ultimately violent purpose. Test 43 in September 1957 was the only surface burst on Novaia Zemlia at the Chernaia Guba site. It was used to study the impact of nuclear blasts on ships and shoreline structures, and in this way resembled the US Operation Crossroads tests at Bikini Atoll beginning in summer 1946. Test 43 was set off on a 15-meter tower 100 meters from the shore with a power of 32 kt. Seven ships and five submarines were anchored at six radiuses: 300, 600, 900, 1,500, 1,900, and 2,200 meters. Yet after all the careful planning, the device failed to ignite because fuses on the receiver and transmitter had blown. The authorities sent in a group of "daredevils" to fix the fuses, and the test proceeded without another hitch. (It is worth noting that, in the Soviet nuclear enterprise, this kind of unthinking heroism, or downplaying of risks, or not showing fear of radiation and so on was commonplace as if a sign of real masculine courage.) In any event, the airborne shockwave, not waves, virtually destroyed the ships which were also completed contaminated with radiation, and sea water and earth thrown into the atmosphere covered most of the South Island, and as far away as 1,500 km to the southeast.[31]

In October 1954, ten to thirteen battalions of military construction workers with their equipment arrived in the area of Chernaia Guba at nuclear test site, located on the southwestern coast of South Island. Their main purpose was to build the auxiliary facilities necessary for the first Soviet underwater tests. Between 1955 and 1962 two above-water nuclear explosions were conducted as were three underwater tests. Later, wells were drilled 20–25 km west of the original test site to conduct underground nuclear tests. The above-ground explosion was carried out on September 7, 1957. The 32 kT charge was on a tower 15 meters high, located 100 meters from the shore. The resulting crater was 80 meters in diameter and 15 meters deep. The test released significant radioactive contamination. One hour after the explosion, the intensity of gamma radiation near the epicenter was 40,000 X-rays per hour. A second highly contaminated zone was formed as a result of an above-water explosion on October 27, 1961, 6 km east of the bay. Both areas are still considered contaminated and access is prohibited. The total area of these contaminated zones is about 100 square kilometers.[32]

Tsar Bomba (so called project 602) was a three-stage hydrogen bomb that yielded about 50 megatons or ten times the amount of all the explosives used in the Second World War combined. The Soviets started developing Tsar Bomba in 1955 in Snezhinsk. The project was transferred with about three months to go before testing to Arzamas-16 where even Sakharov took part in its design. One idea was to detonate a 1.5 megaton plutonium charge to ignite fusion and yield 100 megatons. They eventually determined this was too much even for a deserted Arctic site because of the enormous amount of radiation that would be released and inserted a lead cladding instead of a second uranium cladding, thereby reducing the power of the bomb by about half.

Moving the bomb by train to the Murmansk peninsula, they loaded it onto a Tu-95B bomber that took off toward Sukhoi Nos accompanied by a Tu-16A laboratory aircraft to take videos. Two hours after takeoff, the bomb was dropped by parachute from an altitude of 10,500 meters. It detonated barometrically at 4,200 meters above sea level. The Tu-95B had flown away 39 km by the time of the explosion, and the flying laboratory was 53.5 km away. The radius of the fireball grew to 4.6 km, and might have reached the earth's surface except for the reflected shock wave that threw it upward, and the height of the "mushroom" topped 67 km with a diameter reaching 95 km. The radiation could potentially have caused third-degree burns up to 100 kilometers away. The seismic wave triggered by the explosion circled the globe three times. The shock wave caught up with the Tu-95B at 115 km, but did not endanger the aircraft; the laboratory airplane was 205 km away from the explosion site by the time the shockwave hit.[33] The massive weapon weighed 26 tons and could hardly be transported to a target; another such huge device was never produced for nuclear war.[34] The measured yield of 58.6 megatons exceeded the design power of 51.5 megatons by almost 15 percent, indicating that weapons of mass destruction can hardly be controlled and used in any kind of rational way. Ionization of the atmosphere caused radio communication interference hundreds of kilometers from the test site for about 40 minutes. And the sound wave reached Dixon Island of about 800 kilometers away.[35] A North Fleet officer recalled,

On the day of the next test at the set time I was behind the building with my back to the explosion site. The bright sunny day faded away as if it did not exist at all. A white all-consuming flash covered everything around me for an incomprehensible distance. The earth shuddered, the sound went to the Ural Mountains and back for a long time. What I saw froze my brain, my body went numb, my eyes refused to accept what was happening as reality. Nothing like this had ever happened on Earth in its entire history.[36]

Because of the extent of radiation damage, specialists were permitted to examine the crater and physical destruction only several months later.

By the end of the 1950s, the superpowers were trying to agree on mutual disarmament. However, neither negotiations directly between the leaders of the USSR and the United States, nor discussions of the issue at the 14th and 15th sessions of the UN General Assembly (1959–60) yielded results, and indeed the United States remained suspicious that the USSR would somehow cheat on any agreement. Khrushchev also demanded the demilitarization of West Berlin. Planned talks at a Paris conference in May 1960 were derailed by the U-2 incident. In the summer of 1961 conflict between the United States and USSR escalated with the construction of the Berlin Wall and the invasion of Cuba by American troops. All of this led the Soviet government to order the resumption of nuclear weapons tests on August 31, 1961. In

Figure 5 An artist's concept of a Soviet SS-25 mobile intercontinental ballistic missile being launched. Developed in the 1970s, the SS-25 entered service in the 1980s. © Wikimedia Commons (public domain).

September Khrushchev said, "Let those who dream of new aggression know that we will have a bomb equal in power to 100 million tons of trinitrotoluene, that we already have such a bomb and we only need to test an explosive device for it." Russian hawks in the twenty-first century claim that the Tsar Bomba saved the world from a new war since the United States would obviously be afraid to engage such a powerful adversary after failed arms talks.[37]

Indeed, Tsar Bomba remains an important artifact in Russia today, a major attraction in a Manege Museum exhibition in 2015, and it is now possible to see a previously secret Soviet documentary of the detonation; Rosatom released a film clip in honor of the seventy-fifth anniversary of the Russian nuclear industry. The film shows the main stages in the creation, transportation, and actual detonation of the device along with the claim that in spite of the radiation it showered on the Northern Hemisphere it "is considered to be one of the cleanest ever detonated."[38]

Nuclear Strategy

Soviet leaders sought a ban on the spread of nuclear weapons longer than the United States in a series of actions and formal documents dating to the late 1940s. The Eisenhower administration surely wished to prevent proliferation beyond the three states which held nuclear weapons by 1952 (US, USSR, and UK). But several concerns kept the United States from working toward an agreement. The first was that the Soviets would not agree to onsite verification or inspection, and national technical means available to verify Soviet compliance—spy planes and seismographs, later satellites—were not yet sufficient for the task. The second was that conservative American policy makers refused to consider a ban on the use of nuclear weapons in defense of Western Europe. And both countries believed that continuing to build their own weapons and delivery systems was the only obstacle to preventing first-use of weapons through "mutually assured destruction," or MAD.[39]

From the mid-1950s developments in military doctrine accompanied official Soviet efforts to achieve some sort of international agreement. The

Soviet representative transmitted a non-proliferation provision to the UN General Assembly in September 1957, and the USSR supported the Irish General Assembly resolution opposing proliferation in 1958. The USSR adopted a unilateral moratorium on detonations in March of 1958, and it sought nuclear-free zones in the globe, geographically defined regions in which nuclear weapons would be prohibited. Yet if Stalin had been unsure of their value, Soviet military leaders came to view nuclear weapons as suitable for general war. By the late 1950s they had advanced the strategy that "focused on apocalyptic scenarios for fighting a world war with nuclear weapons and stressed the need for mass armies. The idea of preemption resurfaced, this time on an intercontinental basis, because the Soviet Union had acquired nuclear intercontinental ballistic missiles (ICBMs) and could threaten the territory of the United States."[40]

At the same time, Soviet military thinkers recognized the importance of surprise in the initial period of war, and of using nuclear strikes to determine the outcome of war. In February 1955 in *Voennaia mysl'* (Military Thought) Marshal Pavel Rotmistrov published an article in which he suggested the importance of landing the first, "preemptive" nuclear blow to destroy the enemy's weapons when the latter was preparing a surprise attack. Other military thinkers believed that they could achieve victory "by delivering preemptive nuclear strikes on objectives deep in the enemy's rear and, subsequently, by encircling, cutting off, and destroying the enemy's troops with nuclear and conventional munitions." Yet they also recognized that radioactive contamination, fires, and floods caused by nuclear strikes "could interfere with the success of operations." Simultaneously, scientific leaders began publicly to attack atmospheric weapons tests as a huge risk to public health worldwide because of fallout.[41]

The Military Atom Rivals the Peaceful One

The Soviets were thus certainly capable of engaging the capitalist west, and in particular the United States, in the arms race, in developing nuclear weapons, in pursuing the full range of potential armaments, from strategic to conventional nuclear devices, of increasing yields, adjusting the mix of blast, heat, and radiation to achieve particular military ends, of delivering warheads on jet bombers, then ballistic and cruise missiles, and from submarines, and of building them in factory conditions, scores at a time. Soon after the explosions of the first Soviet atomic bomb in 1949 and hydrogen bomb in 1953, the Soviet armed forces acquired nuclear weapons. Computers and automated control systems followed in an ongoing society-wide scientific and technical revolution. And when Stalin, who had downplayed the importance of nuclear weapons, died in 1953, the military engaged a revolution in strategy, too, that would include nuclear weapons.

Another major feature of Soviet engineering was gigantomania combined with hubris: big projects and big devices and so-called objects to bring together large numbers of people in national campaigns. From the Turkestan-Siberian Railroad (Turksib) of the late 1920s, to the Dnieprstroi hydropower station to the Magnitogorsk steel mills and on and on, Soviet leaders and planners pursued economic development through large-scale technological systems. Not surprisingly, the Soviet nuclear program also was gigantic in its pursuit of uranium ore and nuclear fuel to power the military and civilian enterprises, in weapons manufacture and reactor construction in larger and larger standardized designs. The epitome of weapons of mass destruction was Tsar Bomba.

In the years leading up to the explosion of Tsar Bomba, the Soviet nuclear establishment had served Nikita Khrushchev's foreign policy efforts to achieve some kind of arms control, yet be ready for nuclear war. Khrushchev advanced the doctrine of "peaceful coexistence" versus the Stalin's "hostile capitalist encirclement" that required preparation for inevitable war. In peaceful coexistence, the superior Soviet system would inevitably win in the struggle capitalist in the economic, social and cultural spheres. Yet the Red Army leadership and Khrushchev himself also believed in the need to achieve parity with the United States at any cost. Still, Khrushchev pursued arms control, and Tsar Bomba had a perverse role in this. It was likely ready for testing in 1959, but Khrushchev, seeking some kind of international arms agreement and hoping to improve relations with the United States, ordered a delay in the launch and a unilateral moratorium in weapons tests. Khrushchev wanted to celebrate—for domestic and international propaganda purposes—the "peaceful" atom. In January 1960, Khrushchev unveiled a new nuclear strategy in a speech to the Supreme Soviet. According to Khrushchev, this strategy's aim was deterring war rather than fighting it. But despite Khrushchev's emphasis on deterrence and reductions in military manpower, the nation had embarked fully on the arms race, distrusting American intensions, fearful of another world war, and determined to emerge from the Cold War victorious. Toward that end, the nation simultaneously engaged a large-scale program for peaceful nuclear applications.

CHAPTER 3
NUCLEAR PEACE

In 1954 in Obninsk, Kaluga province, on the shores of the Protva River, in the newly founded Institute of Physics and Power Engineering, in a non-descript factory-like brick building, the AM-1 ("Атом Мирный," or Peaceful Atom) reactor, at 5 MWe and 30 MWt, in many ways an early version of the Chernobyl-type RBMK reactor, came on line, producing the world's first nuclear power for a civilian electrical grid. The Soviets were first with the peaceful atom! The remarkable startup occurred less than a year after President Eisenhower's "Atoms for Peace" speech to the United Nations (December 1953) that called for international control of fissile materials and promotion of applications in industry, agriculture, medicine, and especially electrical power production. Physicists began construction of the AM-1 in 1951, and thus were prepared for the propaganda and ideological victory to assert being the first in the world of the peaceful atom. The AM-1, and the launching of the Sputnik satellite three years later in October 1957, signaled to many citizens and officials that the Soviet Union had entered an age of communist constructivism: the nation would win the economic and political battle against capitalism on the foundation of modern science and technology. Stalin was dead, the glorious future was upon the socialist world.

From within the walls of Obninsk and a dozen other newly founded institutes and science cities of the USSR peaceful applications were pushed by scientists, officials, and journalists alike. They entranced the public with the hope that the future had arrived, a future of an endless supply of electricity, nuclear powered locomotives carrying passengers and freight along the Trans-Siberian railroad, food irradiation that prolonged shelf life of foods and might even lead to shorter lines in stores, and so on. They touted these applications in the central

Figure 6 A Soviet Postage Stamp (1955) celebrating the first-in-the-world reactor to deliver power to the civilian grid, in Obninsk, Russia. © Wikimedia Commons (public domain).

press and republican papers; on radio and television; and in such journals as *Nauka i Zhizn'* [Science and Life], *Tekhnika—Molodezhi* [Technology—to the Youth!], *Ogonek* [The Little Light]; *Vokrug Sveta* [Around the World]; in educational magazines; in workers and peasants' newspapers; and in the "Red Corners" (propaganda points) and on the bulletin board newspapers (*stengazety*) of all sorts of establishments from Kaluga to Kostroma to Komsomolsk-na-Amure. From Obninsk itself as a symbol of unlimited hope to use nuclear energy to heat and power Arctic conquest, Siberian oil development, and Central Asian agriculture, they rolled out a portable nuclear power station that moved about on tank treads.

Igor Kurchatov took a central role in advancing various peaceful nuclear utopias. His deputy, Nikolai Golovin, told me how, after successful detonation of the bomb had been carried out, Kurchatov pushed for a national nuclear electrification program, no longer, apparently, wishing to be connected solely with WMD. He worked with leading scientists to bring the AM-1 on line, all the while thinking of other reactor prototypes. In his 1956 twentieth Party Congress speech Kurchatov presented an outline program for reactor development, leading to the first industrial reactor at Beloiarsk station in the Urals mountains by the early 1960s. Taken together, these hopeful, utopian visions might be called "nuclear peace." They had physical manifestations, but also important cultural, political, and economic meanings. Nuclear peace provided the context for the development of new nuclear applications, some of them far-fetched, yet all of them by and large pursued avidly by hubristic scientists before an awed public.

Nuclear peace was part of a broader postwar "cult of science" that paved the way for expansion of the atomic enterprise and its programs. In this cult, initiated by postwar successes in nuclear weapons and rocketry, politicians, specialists, and the public came to see physics as key to solving a broad range of problems, including economic ones. Nourished by the de-Stalinization thaw, scientists reestablished firmer control over the direction of research and of the scientific enterprise generally, rebuffing interference in their work of Stalinist ideologues. Khrushchev rejected Stalin's so-called cult of personality and the destruction of Leninist norms of behavior among comrades that led to terror in society and within the party itself. Khrushchev's thaw saw reforms beyond politics in literature, education, and science. Khrushchev embraced scientific achievements as confirmation of his own successful rule. He believed science was the key to building a communist society in short order. Toward that end, he supported the rapid expansion of the scientific enterprise in numbers of institutes and scientists. He endorsed the creation of a series of relatively autonomous research campuses or cities, notably Akademogorodok in Siberia.

Nuclear power and space research stood at the center of the cult of science, accompanied by an attitude that such large-scale, expensive and highly visible projects as nuclear fission and fusion reactors, particle accelerators, high-voltage power lines, and spaceships should be at the center of national programs. Space and nuclear power had great ideological significance as well, both before the country as proof that the nation had "reached and surpassed" the West and before the West that the USSR was a scientific superpower in its own rights.

Postwar Atomic Culture

A special kind of atomic culture developed from the mid-1950s. It had physical and political ideological components. The physical components concerned successes in nuclear weapons—meeting the militaristic threat of the United States by detonating the first Soviet atomic bomb in August 1949—and in peaceful applications, especially in nuclear power. In general, by the mid-1950s physical achievements left the secret sphere of weaponry and entered the public sphere of promised applications in industry, agriculture, medicine, and energy.

The ideological component of atomic culture was connected with the celebration of the achievements of physicists and the accompanying increasing internationalization of science. Under Stalin, scientists had lived and worked in an autarkic system where they had to be wary of recognizing the primacy of western specialists or embracing philosophical ideas that Stalinist ideologues claimed were dangerous to the USSR, for example, several epistemological conclusions that emerged from relativity theory and quantum mechanics. They were denied the chance to travel abroad, even to attend international conferences, let alone to share recent publications through the mails, lest they be accused of sharing state secrets. After the death of Stalin, however, the party increasingly recognized the importance of engaging in international science, and encouraged joint research efforts, in particular with East European socialist specialists. Leading scientists and officials held the hope that Soviet scientists would demonstrate priority of discovery in a number of fields. At the same time, scientists pushed to end autarky by asserting authority in science over ideologues. A symbolic step in this public celebration of Soviet science was the effort of such physicists as future Nobel laureate Vitali Ginzburg, himself a participant in hydrogen bomb research, to celebrate the fiftieth anniversary of Albert Einstein's first article on special relativity (1905) in 1955 at the prestigious Academy of Sciences—Einstein's relativity and Nils Bohr's indeterminacy in quantum mechanics had been wrongly equated with philosophical relativism among many ideologues. Scientists announced their primacy in these matters officially in 1958 at a national conference in Moscow on philosophy and modern science. Their achievements in the peaceful and military atom had given them the authority to do so.

More to the point, throughout the 1950s and 1960s scientists joined journalists to celebrate the peaceful atom in such leading party newspapers as *Pravda* and *Izvestiia*, in weeklies, in literary journals, in local and regional publications and other venues. While one important theme of the articles was the assertion that the prevailing applications in the United States were military to reflect capitalism's warlike nature and the interest of business monopolies, not those of the workers, the vast majority of articles touted the myriad peaceful applications that had found full radiance in the USSR. Popularizers stressed applications in industry, agriculture, medicine, and transport. Radioactive tracers in agriculture were a magic wand of alchemical possibilities that would rework plant physiology and fertilizers (a new technology would apparently overcome long underperforming collective farms without needed to reform them). Food irradiation would increase shelf life of a variety of products, conquering muddy roads and the absence of refrigeration equipment in the USSR that saw high spoilage rates. Some of the proposed possibilities seem far-fetched, not only in retrospect: nuclear-powered automobiles, locomotives, icebreakers, submarines, and jets, the latter two to be sure of military significance. (The United States also spent billions of dollars these programs—rocket ships and jets and locomotives—*also* without significant results.) Finally, the Soviets touted peaceful nuclear explosions (PNEs) to correct the "mistakes" of nature, fill deserts with water, build dams, and several other massive excavation projects.[1]

Knowledge of atomic energy entered the public domain through well-attended popular science and science courses on nuclear energy and nuclear physics and lecture series. Journalists called for citizens to be equipped with Geiger counters to help in prospecting for uranium ore. They urged the public to write scientists with suggestions for applications. The archives of the Kurchatov Institute for Atomic Energy contain letters from ordinary folk to Kurchatov himself promoting a variety of nuclear potential projects; one suggested portable nuclear generators to power higher productivity in agricultural fields. The popularization effort fit with ongoing campaigns to demonstrate that the USSR had reached a new, and perhaps final stage on the path to communism, based on the achievements of modern science and technology. An ongoing and powerful "scientific-technological revolution" was advancing all areas of the economy forward, with energy production only one area of rapid change. Party leaders had chosen the correct path for "constructing material basis for communism" in which nuclear power would play a major role. As for young people, the press was filled with articles and stories; cartoons and drawings; children's rhymes and poems; poetry and prose in a variety of literary journals (*Prostor*, *Smena* and *Neva*, among others) and bold editorials.[2]

Nikita Khrushchev Embraces the Atom

Nikita Khrushchev and Igor Kurchatov contributed to the cult of the atom directly through public acknowledgment at home and the sharing abroad of some of the stunning Soviet achievements. In August 1955 Kurchatov and other Soviet delegates astounded the attendees of the first International Conference on the Peaceful Uses of Atomic Energy in Geneva, Switzerland, with the presentations on the Soviet peaceful atom, especially the work of Igor Tamm and Andrei Sakharov on controlled thermonuclear synthesis (fusion, dating to 1951). The Soviets arrived with the largest delegation, larger than that of the United States among attending countries, with roughly 1,000 individuals who presented seventy-nine papers and put up exhibitions of some of the potential industrial, agricultural, and other applications.[3] In all, the first Geneva conference was, as participants noted, an outstanding scientific success, even if they learned nothing deeply new from other nations. Rather, they had learned that their scientific colleagues' interests were similar to theirs. In scores of sessions over two weeks they made acquaintances, lifted the veil of secrecy, created a truly international setting to exchange ideas, and gauged the vitality of others' research programs. The enthusiastic response to the appearance of Soviet specialists at the conference was captured in the first fifteen pages of the October 1955 issue of *Bulletin of the Atomic Scientists* that praised the end of secrecy and beginnings of hopeful cooperation between Soviet and western physicists.[4] The second Geneva conference held over two weeks in September 1958 produced over thirty volumes of papers and repeated calls for world collaboration.

Building on the momentum of Geneva, in February 1956 Igor Kurchatov appeared at the Twentieth Congress of the Communist Party. Best known for Nikita Khrushchev's so-called secret speech that condemned the murderous excesses of Stalinism and for triggering the de-Stalinization thaw, Kurchatov's speech also provoked wonderment. In his presentation, Kurchatov proposed an aggressive program for the commercialization of atomic energy and applications in industry, agriculture, and medicine—that indeed were adopted in Russian industry over the next half century. Kurchatov pointed first to the rapid growth of the powerful army of young scientists, engineers, and designers who were prepared to solve problems of the nascent industry. In specific, he outlined how in the near future they would bring on line 2 to 2.5 GW of atomic energy, including a 1 million kW unit in the Urals and a 400,000 kW station closer to Moscow. These were industrial prototypes that would set the stage for the full flowering of nuclear energetics. On top of this, he suggested that the industry would build up to ten different kinds of reactors at 50,000 to 200,000 kW: fast reactors and slow, with graphite, beryllium, heavy and simple water moderators, and with gas, water, and liquid metal coolant. He acknowledged the higher capital costs for nuclear energy than for coal powered plants, but noted uranium fuel had volumetric advantages. He turned to applications in transport where, within the next few years icebreakers and other ships that could sail up to three years without refueling, would be built. Kurchatov concluded his speech talking about fusion power. He called for joint research, including with American scholars, to solve these important problems, and asserted that all that was required was for the US government to agree to outlaw nuclear weapons for collaborative work to proceed.[5]

If Kurchatov's contribution to the atomic bomb project had been a secret, now it was clear that he was the leader of the nuclear enterprise. With Khrushchev in April 1956 he visited Harwell, England, and the UK's Atomic Energy Research Establishment. The trip signaled the determination of Soviet physicists to participate in the worldwide development of nuclear power. "The Beard," as Kurchatov was known, gave a talk before 300 specialists on the possibilities of fusion—controlled thermonuclear synthesis—that encouraged many of the attendees to think that cooperation between the superpowers and nuclear nations in a variety of areas would be possible. Until his premature death in 1960, Kurchatov played a major role in pushing cooperation, inviting "capitalist" scholars to Moscow and beyond to visit with Soviet scholars at their previously top secret facilities, and helping Soviet physicists to undertake research trips and fellowships abroad.[6]

Over the next years Kurchatov and other physicists, engineers, and their allies put nuclear energy on the agenda by pushing a series of arguments about how to pursue nuclear power, and advancing a series of different industrial prototypes to meet the nation's future energy needs that became the industry's standard PWRs, RMBKs, and LMFBRs.[7] First, they noted the geographic imbalance of resources that saw fossil fuels and future hydroelectric potential largely in Siberia, but industry and population in the European Soviet Union. They touted volumetric considerations—a few grams of uranium had the energy potential of tons of coal or diesel fuel. They claimed that through standardization and industrial assimilation they would make nuclear power plants competitive with fossil fuels in terms of capital costs. They proposed cutting down on capital costs by using less steel and less reinforced concrete in construction without lessening reactor integrity. They pushed ever-larger reactor units to achieve economies of scale. And they argued that future fuel demands would be met by creating a fleet of breeder reactors to produce plutonium. Into the pantheon of reactor models, they proposed the AST to produce steam heat for industrial and domestic purposes located within or close to major cities, and even mobile and floating NPPs that have application today. Some of the variants were not very successful. They built a prototype of the Arbus (Arctic Block Unit) of modular components that might be shipped or parachuted to the Far North for assembly.[8]

The nuclear icebreaker fleet developed in parallel with domestic nuclear power.[9] The decision to build the first nuclear icebreaker was taken on November 20, 1953, and six years later the "Lenin" was launched on September 5, 1959. A matter of national pride, and a typical all-union hero project, over 500 manufacturing plants and organizations all over the country were involved in its building. Its sailors relished gleaming white interiors, polished wood railings, sauna, and swimming pool. Other vessels joined the nuclear caravan beginning in the 1960s and 1970s: "Arktika," "Siberia," "Rossiia,"

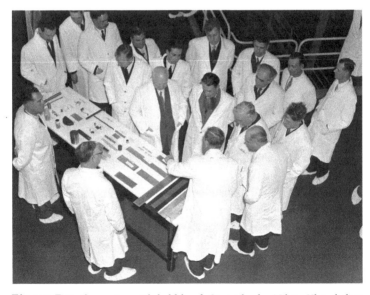

Figure 7 In the center, with bald head, Soviet leader Nikita Khrushchev, and Igor Kurchatov to his left, April 26, 1956, thirty years to the day before the Chernobyl disaster, during a visit to the site of the British nuclear establishment, Harwell, England. Kurchatov surprised his English audience with discussion of Soviet experiments on nuclear fusion. © Wikimedia Commons (public domain).

"Sovetskii Soiuz," and "Iamal" built at the Ordzhonikidze Baltic Plant in Leningrad, and the "Sevmorput'" lighter container ship was built at the Butoma Peninsula Shipyard in Kerch in Crimea that Russian annexed from Ukraine in 2014. The "Taimyr" and "Vaigach" nuclear icebreakers were built at Wärtsilä Marine in Finland (1985–9) using Soviet equipment and steel, and both were launched on the eve of the breakup of the USSR.

Socialist Cooperation

Soviet successes found response in the socialist world in joint research and in domestic nuclear energy programs. In July 1955 specialists from the socialist world gathered in the Central Hall of the newly opened central skyscraper of Moscow State University in an Academy of Sciences session on the peaceful uses of atomic energy. Several of them presented papers they would read again in Geneva in August. They gathered informally in Geneva several times, including at a restaurant on the shore of Lac Leman where they discussed the nearby CERN (European Organization for Nuclear Research) founded in 1954. They envied its 600 MeV synchrontron and 30 GeV proton accelerator, but thought they could do better. They considered the possibility of such a center for the socialist countries. In March 1956 they agreed to establish Joint Institute for Nuclear Research (JINR) in Dubna, Russia.[10]

Initially, the USSR invited China, Poland, Czechoslovakia, the DDR, and Romania to work with Soviet-designed accelerators and experimental reactors, to acquire the nuclear materials needed for research, and promote scientific exchanges as well. They built on bilateral agreements, already signed in 1955 that included Hungary and Bulgaria, and later Yugoslavia and Egypt, through which the Soviet Union would supply the equipment, apparatuses, and know-how relating to those installations, and "assign Soviet specialists to aid in assembling and putting them into operation, to furnish the necessary amount of fissionable and other materials for the piles and for research work, and to deliver the necessary amounts of radioactive isotopes to these countries until such time as the experimental atomic piles received from the USSR were put into operation."[11] The signatories would pay over several years for the expensive equipment and promised to establish the due secrecy and safety regimes. All of this permitted the USSR to claim that its peaceful atomic cooperation preceded that of the United States.

First Steps toward Arms Control

How could the Soviet Union both celebrate its peaceful atomic programs—as the United States had, and ignore the fact that it had fully engaged the United States in an arms race? How might Soviet scientists lobby their government to accept arms control endeavors? Given their stature in society, mightn't the emboldened scientists begin to contest continued weapons development, increases in yield and ease of delivery? Surely, reaching an agreement would be easier since, from 1955 at Geneva, US and Soviet scientists had started to become acquainted, understood that the countries were on roughly the same level in scientific expertise and development of both peaceful and military nuclear technologies, and there was unlikely to be a technological breakthrough giving one country a significant military advantage over the other, or so many of them assumed. At the very least, they believed that they could trust one another and advise their governments to move ahead on good faith.

Yet, of course, there were problems in reaching arms control agreements. One was, in fact, the way in which in both nations a number of leading officials and scientists believed that the other side was evil, could not be trusted, and would seek to violate any agreement to take advantage of any agreement. Second, as yet national technical means (seismographs, spy planes, soon satellites) were inadequate to verify compliance with certainty. Third, the United States was insisting on some on-site verification of weapons testing sites which the USSR fully rejected.

Another obstacle was the absence of such channels for Soviet specialists that were available to western scientists to express their worried viewpoints on the crucial policy issue and world danger of the arms race as the Federation of American Scientists (FAS, 1945-present). The FAS involved many scientists connected with the Manhattan project who recognized how their work had ended the war at Hiroshima, but then had misgivings over Nagasaki, and increasingly saw that their role must be one of securing peace.[12] Such scientists as Kurchatov, while supporting weapons development, struggled to shift the nation's focus to peaceful technologies—as his speech at the twentieth party congress indicated. Kurchatov's deputy, Igor Golovin, turned to the study of applied thermonuclear research. Andrei Sakharov who helped build the hydrogen bomb, began to agitate for arms control, and later devoted himself fully to human rights, losing his position at "the Installation." Yet Soviet specialists had no official organizations or scientific societies to push more broadly in that direction, although they were closely involved in the activities of such international organizations as Pugwash and Physicians for Social Responsibility.[13] Indeed, the closed Soviet political system would not permit arms control NGOs to develop outside of strict government control. The autonomy of scientists was always limited.

Sakharov who considered himself a true patriot worked on the atomic bomb project enthusiastically, but by the mid-1950s began to worry about nuclear fallout and the potential for a nuclear war. The USSR exploded at least 718 nuclear bombs between 1949 and 1998, until 1962 to vast majority of them above ground spreading fall out throughout the northern hemisphere.[14] The first Soviet test of a thermonuclear weapon (RDS-37) had killed a number of bystanders standing kilometers away through radiation or shock wave. Sakharov calculated that the non-threshold radiation effects from the redundant weapons tests would kill hundreds of thousands of people over time. He turned to Kurchatov for help in pushing a moratorium of Soviet tests. Kurchatov was very ill at the time recovering from a stroke. Kurchatov quizzed Sakharov and, realizing Sakharov's concerns were valid, flew against the orders of his doctors to Crimea where Khrushchev was vacationing. Khrushchev grew angry, refused Kurchatov's call for a moratorium, and Kurchatov temporarily fell out of favor.

Andrei Sakharov was rewarded for his patriotic contribution to the nation's defense by designing the hydrogen bomb, becoming a doctor of science at only twenty-six years old and a full member of the Academy of Sciences at thirty-two, and being given financial rewards including special housing. The public could only speculate why such a young man had such status because his connection with the hydrogen bomb was a secret. Sakharov himself, originally enthusiastic about the need to counter the American threat, grew to have misgivings about the bomb—or rather that the generals would hardly be circumspect in using nuclear weapons, and he began to voice his discontent about the bomb, eventually becoming a dissident. His feelings were hardened by the test of the RDS-37. From the late 1950s the United States and USSR tested ever larger devices, leading to a significant rise in radioactive fallout and global radiation levels.

By 1957 Sakharov and Kurchatov had begun writing and publishing about the dangers of radioactive fallout, and although focusing on US behavior, it was clear they had in mind the danger of all weapons testing. There could be no clean bombs with less radioactive debris, no matter what facilitators of the arms race as the US pusher of the hydrogen bomb, Edward Teller, tried to insist. Ultimately, Sakharov calculated that a one-megaton "clean" H-bomb would produce enough radioactive carbon to result in 6,600 deaths worldwide over the next 8,000 years. He published these conclusions in the leading Soviet journal of atomic energy (*Atomnaia Energiia*) shortly after the Soviet Union announced a temporary moratorium on nuclear tests. Sakharov wrote,

> The remote consequences of radioactive carbon do not mitigate the moral responsibility to future victims. Only an extreme lack of imagination can let one ignore suffering that occurs out of sight. The conscience of the modern scientist must not make distinctions between the suffering of his contemporaries and the suffering of generations yet unborn.[15]

Sakharov expressed his concerns directly to Khrushchev who rejected them. When, in July 1961, Khrushchev decided to revoke the moratorium on bomb tests, Sakharov objected that a renewal of testing would yield little technically while damaging international security. Still, he followed the orders of the head of state and took part in the initial preparations for testing the Tsar Bomba. According to one estimate, the 1961 Tsar Bomba, an explosion that by itself accounts for about 10 percent to the total yield of all atmospheric nuclear explosions in history, would injure or kill about half a million people over time. Sakharov turned toward gaining approval for the 1963 limited test ban on atmospheric tests.[16]

Ultimately, however, the work of Kurchatov, Sakharov, European, and American physicists bore results. In 1963 the Treaty Banning Nuclear Weapon Tests in the Atmosphere, in Outer Space and under Water (LTBT) was signed by the USSR, the UK, and the United States.[17] Negotiations initially focused on a comprehensive ban, but this hope fell by the wayside because of lingering problems getting agreement on the detection of underground tests and Soviet concerns over the intrusiveness of proposed verification methods. By April 15, 1964, six months after the treaty went into effect, more than 100 states had joined the treaty as signatories and thirty-nine had ratified or acceded to it.[18] Indeed, compliance with the partial test ban treaty has been very good among signatories. Yet nuclear tests, even so-called Peaceful Nuclear Explosions, have led to the accidental release of radioactivity debris into the atmosphere, in part through purposeful (and unavoidable!) "venting" of debris. Underground tests likely lead to long-lived radionuclides, including ^{135}cesium, ^{129}iodine, and plutonium to seep into the ground. An example of this is the Soviet Chagan Lake test (see below) of 1965. There could be no clean or good nuclear weapons. But was nuclear peace at hand?

CHAPTER 4
NUCLEAR HUBRIS

Nuclear Russia reached its pinnacle of power in the 1970s and 1980s under General Secretary of the Communist Party Leonid Brezhnev (1906–82). Brezhnev, who had deposed Khrushchev in 1964 for his "hare-brained schemes," determined to replace Khrushchev's erratic leadership with one based on cooperation and understanding among leading officials ("trust in cadres"). He claimed that the country had entered the stage of "developed socialism," seeking to convey the message that socialist society had transformed into something qualitatively more advanced than in the Stalin and Khrushchev eras, and he claimed it rivaled the capitalist west. Economic growth, progress in culture and science, power and influence abroad—all of these things indicated such achievements.

By the 1970s Soviet leaders, economic planners, and industrial managers had succeeded in their long-term goals of achieving military and economic parity with the West. According to the official statistics, Soviet industrial output had reached 80 percent of that of the United States. The USSR was the global leader in production of coal, iron ore, tractors, cement, coke, cotton, and wool, and it outstripped the United States in annual rates of growth of cast iron, steel, mineral fertilizers, and train locomotives. Gigantic new industrial combines and infrastructural projects (power stations, roads, bridges, power lines, a new trans-Siberia railroad known by its acronym as BAM, and massive hydroelectric power stations) spread from the European USSR into Siberia, the Far East, and Central Asia. Nuclear power was both a symbol, and a crucial component, of these economic and technological achievements.

The Brezhnev era saw, at least in the minds of Communist Party ideologues, the full flowering of socialism. But since Khrushchev had embarrassingly promised that the USSR would reach the full flowering of *communism* by 1980, the Brezhnevites thought up a new stage of glorious Soviet perfection to indicate great social, cultural, and economic progress: developed socialism. In developed socialism, under the enlightened leadership of the party, the economic system would reach full maturity. Factories would produce the mix and quantities of industrial, agricultural, consumer, housing, and other goods long promised, and would outproduce the west in many areas. The Brezhnev leadership advanced a series of major projects to demonstrate "developed socialism": a Food Program (1982) at long last to solve the problem of underproduction of basic foods and increase milk and dairy, meat and chicken, and vegetables substantially (it was a failed program from the start); BAM to speed Siberian settlement and development (BAM was, in fact, completed only twenty years after Brezhnev himself expired); expansion of oil and gas production; efforts at housing; and the construction of increased energy capacity including through nuclear power.

From the founding of the Soviet state, leaders saw electrification as a panacea for the problems of economic weakness, agricultural, and industrial lag and even what they considered to be the cultural backwardness of a largely peasant society. Electrification of the entire country, as the slogan cried out in posters, plus Soviet power, would lead to communism. In the Brezhnev era, planners accelerated efforts to increase electricity production with nuclear power forcefully entering the mix.

There were few hopeful moments in the military sphere. A signal achievement in the nuclear sphere was the establishment of Détente with the United States. Yet Détente hardly slowed the arms race. Détente involved in fact the pursuit of parity in military power that resulted in conventional Warsaw Pact forces having an overwhelming advantage over NATO forces, and the accelerated manufacture and deployment of MIRV-ed (multiple warheads) ICBMs, submarines, and bomber missiles. And the party and military pursued weapons development so single-mindedly as to leave behind dangerous wastes everywhere. Scores of military bases and weapons testing grounds contributed to this toxic legacy through their scale and the criminal mismanagement of ordnance, biological, chemical, and nuclear weapons. The nuclear enterprise may be the worst offender. Since the dawn of the nuclear era, Russia has accumulated hundreds of millions of cubic meters (m^3) and tons of liquid and solid radioactive waste, the lion's share of which comes from the military; this is more than half of radioactive waste accumulated in the world. This waste is stored haphazardly in facilities that are filled above capacity and in poor condition. There are also a large number of not-yet-fully inventoried or even found contaminated sites and waste dumps. Was this developed socialism?

And yet, nuclear wonderment grew in the 1960s and 1970s as part of the society-wide embrace of modern science and technology as panaceas for such problems still facing the socialist motherland as poor consumer goods, inadequate and unwholesome foodstuffs, energy production and efficiency that lagged behind the capitalist west and challenges in harnessing the nation's natural and mineral resources. Nuclear technologies would solve all these problems: food irradiation would ensure that stores were stocked with potatoes and

Figure 8 A typical poster of communist constructivism with the slogan, "The Atom! In service of Peace and Progress!". © Public Domain.

other vegetables in better than a vegetative state; nuclear power would free oil and gas for export to earn hard currency; and the generation of copious amounts of electricity would magically enable the achievement of communism. The state would embark on an aggressive program to build PWRs and RBMKs throughout the European part of the nation—with floating smaller units to power Arctic resource development that appeared out of the Arctic mist ultimately in the 2020s.

Nuclear Fordism

Building from Kurchatov's visionary speech at the twentieth party congress, the nuclear industry accelerated efforts in the Brezhnev era to build scores of reactors, although always falling short of construction targets. The nation would produce electricity and industrial and steam heat from a variety of different reactors: PWRs (again, called "VVERs" in the Russian parlance), at first at 440 MW units, then in 1,000 MW units); the infamous Chernobyl-type channel-graphite reactor (the RBMK, in 1,000 MW units, with two built in Lithuania at 1,500 MW, and with plans to build 2,400 MW, absurdly massive even for Soviet gigantophiles); ASTs on the outskirts of cities in 500 MW and perhaps 1,000 MW, but never built); LMFBRs (the Soviet BN series, with current efforts to standardize at 1,200 MWe units in the 2020s and 2030s); and floating nuclear power stations. In all events, the Soviets claimed that they would build reactors with lower costs than the western experience and within a shorter time frame. The cost and time claims have not been met, and it is likely that nuclear reactors are complex, expensive devices that no one can build cheaply and quickly. The following chart gives a sense of the mix of these reactors in 1982 with total capacity only a bit over 17,000 MWe (see Table 1).

In 2021 total capacity was roughly 29,000 MWe (see Table 2) that produces roughly 20 percent of Russia's electricity. In spite of Kurchatov's vision, only during the 1970s, well after the nuclear take-off in the United States, did the USSR manage a significant industrial effort beginning with the completion of the first 440-MWe

Table 1 Nuclear Power Stations in USSR, 1982

Name	Unit	Gross electrical Power (MW)	Reactor type	Year of reaching nominal power
Novo-Voronezh	I	210	WWER	1964
	II	365	WWER	1970
	III	440	WWER	1972
	IV	440	WWER	1973
	V	1 000	WWER	1981
Beloy arsk	I	100	U.Gr (Chennei BWR)	1967
	II	200	U.Gr (Chennei BWR)	1969
	III	600	FBR	1981
Kolsh	I	440	WWER	1973
	II	440	WWER	1975
	III	440	WWER	1982
Leningrad	I	1000	RBMK	1974
	II	1000	RBMK	1976
	III	1000	RBMK	1980
	IV	1000	RBMK	1981
Armenian	I	407.5	WWER	1979
	II	407.5	WWER	1980
Kursk	I	1000	RBMK	1977
	II	1000	RBMK	1979
Chernobylsk	I	1000	RBMK	1981
	II	1000	RBMK	1981
	III	1000	RBMK	1982
Rovno	I	440	WWER	1981/82
	II	440	WWER	1981/82

Name	Unit	Gross electrical Power (MW)	Reactor type	Year of reaching nominal power
Inzhno. Ukrainskaya (South Ukraine)	I	1000	WWER	1981/82
Smolenskaya	I	1000	RBMK	1981/82
Total	26	17 370		

Several small prototype, and district-heating reactors (VK-50, BOR-60, BN-350, Bilibin, etc.,) with a total capacity of some 900 MV have not been included in this list.

Source: V. A. Semenov, "Nuclear Power in the Soviet Union," *IAEA Bulletin*, vol. 25, no. 2 (1983): p. 48. See also William Davey, *Nuclear Power in the Soviet Block*, Report 8039, (Los Alamos National Laboratory, March 1982), pp. 10–11.

PWR at Novovoronezh in 1972 and the first 1000-MWe RBMK at Leningrad in 1974. There were successes and failures. Some of the more fascinating efforts for PWRs were the Kola NPP, the first station north of the Arctic Circle, with its three reactors crucial to exploiting rich mineral resources on the Kola Peninsula; it includes a salmon and sturgeon fishery with warmer water for the fish coming from the reactor cooling effluent. The BN-600 in Beloiarsk ostensibly came on line in 1981, but was handicapped by frequent and serious accidents and failures. Intensive construction of RBMKs and VVERs followed until the Chernobyl disaster of 1986. The Soviets would sell the technology to the socialist world—the German Democratic Republic, Bulgaria, Hungary, and Czechoslovakia, with Romania opting for Canadian "CANDU" reactors—which they willingly embraced as a sign of modernity. Efforts to sell the technology abroad failed, with the exception of Finland, because the first generations of Soviet reactors had no containment vessel to keep products of fission within the reactor hall in case of an accident; the Soviets used containment only later on when forced by the sale to Finland to do so; and the RMBK had no containment, nor is any possible.[1] In fact, plant construction lagged both in number of new starts and pace, although in the early

1980s officials remained confident of meeting bold targets. By the end of 1982, the total installed capacity of nuclear power plants in the USSR was only 18 GW; these stations generated 86 billion kWh electricity—6.5 percent of the country's total electricity production in 1981. On the eve of the Chernobyl disaster, industry officials claimed generation would reach 220 billion kWh in 1985, or almost 14 percent of the total—and 24 percent in the European region.

One of the ways in which the Soviets—and now the rejuvenated Russian industry—intended to build more rapidly and at lower cost than global experience was through the construction of central design institutes, construction bureaus, and factories that could product component parts and even reactor vessels in a kind of serial production. If submarines could be produced in number, why not reactors? Anatoli Aleksandrov, a leading specialist at the Kurchatov Institute, later its director and president of the Soviet Academy of Sciences, and the moving force behind the RBMK reactor, for which he had no remorse, pushed serial production. With Aleksandrov's help, the Atommash factory was erected in Volgodonsk to take advantage of copious amounts of hydroelectricity and the Tsimilianskoe Reservoir, to float the massively heavy reactor vessels and associated equipment, produced in serial form, to sites far from the Volga River basin. The vision of mass production of reactor equipment, in a new hero city, Volgodonsk, in a new hero factory, Atommash, was not realized. There were significant production delays and a main foundry wall collapsed which suggests that it was premature to build such a factory to produce reactor equipment if they could not build a wall.

The LFMBR is another important mainstay of the Russian industry. Breeder reactors produce plutonium that has immediate military uses and therefore are a great risk to proliferation. The plutonium produced could also be used to fuel other breeders to produce electricity and, when used in combination with mixed uranium oxide fuel, would ensure the long-term availability of nuclear fuel at ostensibly lower cost. But precisely because of proliferation concerns, the United States abandoned the LMFBR in the 1970s, and because of technical problems, accidents, and other concerns the two other leading breeder countries, France and Japan, gave up their plutonium-producing ghosts, leaving Russia as the world leader and only true proponent. The Russian program has been waylaid by high costs, and also serious accidents, sodium fires, and leaks.

The Soviets pursued nuclear power for a variety of reasons, as noted. One of the most important was that about 80 percent of its energy resources were concentrated in eastern regions of the country, while 75 percent of the population and consumers of power were concentrated in the European part of the USSR. The transport of fuel from the east of the Soviet Union to western regions constituted roughly 40 percent of the country's rail freight. In this light, nuclear power was more environmentally sound than fossil fuel power. From the point of view of air pollution and particulate, acid rain, and industrial safety, this was certainly an appealing and reasonable argument. Planners began boldly to forecast new capacity to reach 50 GW by 2000 and 60 to 80 GW by 2010. This required annual commissioning of 8–15 GW capacity, a rate never achieved, based on radical modernization of equipment and its standardization. In all, a nuclear utopia prevailed, from the geopolitical meaning of nuclear weapons and reactors, to economic reasons to pursue nuclear power including a geographic distribution fossil fuel resources far from population centers, to a desire to earn hard currency for oil and gas exports, as well as the cultural significance of the atom as the beacon of the nation's radiant future at such reactor "parks" as Chernobyl NPP.

Indigenes and Nuclear Power in the USSR

Throughout the globe indigenes have paid the price for the development of the atom in terms of their health and destruction of their homelands. Navajo miners in the American southwest saw their homes destroyed and suffered high rates of lung and other forms of cancer for their service to the United States in providing uranium ore to fuel the arms race. Tahitians and Algerians lost their lands to the French for testing bombs, as did Aborigines in Australia, Bikinians to the Americans in the Pacific, Nenets in the Russian Far North and

Kazakhs in Kazakhstan.[2] Chukchis in the Russian Far North were given a "gift" of the peaceful atom—a small nuclear power station based on four reactors of the RBMK design. In exchange for the gift, the Chukchi were expected to thank the Russian scientific civilizers for modern science and medicine, for changing their nomadic reindeer herding into sedentary collective farming and for introducing large-scale extractive industries that caused significant damage to land and water, flora and fauna.

Like the homelands of other Indigenes, Chukotka has been crucial to Russian wealth generation since the empire spread to the Far East and Far North in past centuries. In the 1930s when geologists discovered tin, gold, and other minerals in the region, the Kremlin leadership decided to tap that wealth—and the reindeer, game, and fish of the Chukchi people—by force. They employed the massive Dalstroi gulag operation with tens of thousands of slave laborers working and dying in mines, and collectivizing reindeer herding in order to tie the herders into planned economic activities. (The influx of settlers transformed the Chukchi into a minority people in their own lands; as recently as 2011, of the 50,800 inhabitants of Chukotka only 16,000 were indigenes.) Under Stalin Party officials decided to combat viciously what they perceived to be the Chukchis' backwardness and anti-Soviet behavior. After the death of Stalin, coercion could no

Figure 9 Bilibino Nuclear Power Station, Chukotka, Russia, 150 km above the Arctic Circle, powered by four EPG-6 Graphite Moderated Reactors. Undergoing Decommissioning as of 2021. © Public domain.

longer be the motive force. But the atom might be. Hence engineers built the Bilibino NPP to power the regional mining activities. The Soviets celebrated the construction of the Bilibino Nuclear Power Station, the northern most nuclear power plant in the world, well above the Arctic circle, as part of its effort to create a nuclear-powered Arctic, and to bring a special kind of "fire" to the simple Chukchi people. Commissioned in 1974, the four graphite moderated EPG-6 reactors each produced 12 MW electric and 62 MW thermal power.[3]

The collapse of the USSR was a terrible shock to Chukotka. A deep depression led to significant outmigration of non-indigenous and mostly younger people, with the population shrinking to a third of its former size, while the indigenes have suffered high rates of poverty, unemployment, alcoholism, suicide, and a variety of infectious diseases.[4] Rosatom and Russian leaders determined that they could decommission the Bilibino NPP and replace it with a floating nuclear power station, the "Lomonosov," to power renewed economic growth, and perhaps end these problems.

Yet if Soviet engineers claimed Bilibino was for the Chukchi, in reality it was to power mining, especially gold and tin industries, and hence benefitted Russians; of 5,000 people in the town of Bilibino only 300 were Chukchi, mostly in service positions, and the rest are Russians. And most of the employees except in secretarial positions are male. Marina Kirii, an information specialist at the station, won the annual "Miss Atom" contest, dedicated to demonstrating the soft, feminine side of nuclear power in 2011. Also a homemaker, she "raises a beautiful daughter. And she gave her prize—a tour to Brazil, or rather, the money that was to be spent on the prize—to the charity foundation Chulpan Khamatov 'Gift of Life' foundation to help children with serious illnesses. Thank you, Marina, for your beauty, for your optimism, for your work for the benefit of the nuclear industry, and for your kind heart."[5] The surrounding environment has accumulated radioactive contamination sediments, soils, lichen, moss, and reindeer meat with the impact of anthropogenic radionuclides on Chukchis about ten times higher than that of non-indigenous residents.[6]

The "Academician Lomonosov" floating nuclear power station, with two 35 MWe reactors like those of the icebreaker fleet of the

KLT-40S type, moored in Pevek on the Arctic shore, began operation in 2021 to replace the outdated Bilibino station.[7] Some analysts worry that the barge that holds the two nuclear power units are vulnerable to tsunamis and other violent sea surges that could waterlog its reactors. Meanwhile Rosatom officials claim the floating plant is steeled against such calamities, citing that the fact that it has 24 hours of backup coolant should its reactors endure a Fukushima-like inundation. Yet there is also the question how to remove and store SNF is such a difficult climate and remote location. Another question is whether the Lomonosov will be used for Bilibino, or rather will power offshore oil drilling operations in the Chukchi Sea. Most indications are that the floating atom is for oil and gas, not for the economically depressed interior and a dying town, and this will lead to further economic hardships in the city and its surroundings.[8]

PNEs

One example of the fetishism of Soviet nuclear bomb builders was an extensive Peaceful Nuclear Explosions (PNE) program. Visionary physicists in the United States also pursued the idea of turning "swords into plowshares" (the US "Project Plowshares"[9]). The idea involved making the atom less frightening to the public, demonstrating to Cold War opponents that one's research and development program was on target to engage on any level, and finding economic value in the nuclear explosions. It was also to continue to test nuclear devices after the signing of the LTBT. Indeed, Leonid Brezhnev ordered the acceleration of the Soviet program in the 1960s at the time of increasing tension between the superpowers to demonstrate the technical skills of Soviet weapons designers and to contribute to the national economy. The Soviets conducted over 120 PNEs, 116 from 1965-1988, many at the 5–10 kiloton range, but also at 10–20 kilotons and even larger, with some local venting and pollution.[10] What couldn't nuclear weapons do? They could put out runaway oil well fires, build underground storage caverns for hazardous waste, in a few moments create an earthen dam. Among the most ambitious projects considered were a massive on-again, off-again waterworks that included a huge transfer canal to bring Siberian river water to Central Asian orchards with excavation of the canal through dozens of nuclear charges and the actual nuclear construction of Lake Chagan.

Lake Chagan was excavated with a 140-kt underground nuclear explosion, equivalent to 140,000 tons of TNT, on January 15, 1965. Chagan was the first and largest of the Soviet PNEs for earth-moving purposes, on the dry bed of the Chagan River, a tributary of the Irtysh River, selected to lift thousands of tons of riff-raff to dam the river during the spring freshet and create a reservoir. The crater formed by the Chagan explosion had a diameter of 408 meters and a depth of 100 meters. The lake was an abject failure in all ways owing to the intense radioactive residues and rocks left behind, although to this day Russian newspapers report it as being a success. The authorities claimed that after fifty days the lake's radioactivity had lowered to safe levels! Yet roughly 20 percent of the radioactive fission particles released by the explosion escaped into the atmosphere, and some of it was detected as far away as Japan. Fifty years later the lake's ten billion liters of water had one hundred times more than the permitted level of radionuclides in drinking water, and radioactivity reached background levels only 150 meters from the shore.[11] An ecological study spanning from 1981 to 1991 described a two times higher incidence of childhood cancer among those living 200 km from the test compared to those living 400 km away. There are also cytogenetic abnormalities and such genetic aberrations such as chromosomal translocations, satellites, and micronuclei in the descendants of those living close to the testing site.[12]

While designed to be of low radiation and hence "clean," and in principle not to vent, many of the 120 PNEs in the former Soviet Union have in fact polluted the environment, and with all of the other radioactive waste from fallout and haphazard dumping, have had a significant impact on the Arctic region.[13] PNEs conducted 30 km from Kirovsk in the Khibiny Mountain range indicate both the hopefulness of the physicists, engineers, and planners to find economic value in nuclear explosions and the great risks and dangers they must ignore when they justify bold and explosive experiments. The Mining Institute of the Kola Branch of the Academy of Science carried out long studies promoting an experimental nuclear explosion at the mines of the Apatit combine in Kirovsk, itself built with prison labor in the 1930s, with the apatite providing phosphorus for agricultural and other purposes. Mining institute personnel joined those in Snezhensk to design the PNE. A first explosion was conducted in September 1972, followed by another in 1984. The operations, designated Dnieper-1 and Dnieper-2, were carried out in utmost secrecy. The explosions were powerful enough to pulverize the ore into fine granules and dust—thus sparing the country additional expenditures required for ore grinding. In 1974, the Kirovsk Works Combine established a Radiation Safety Service to convince the miners through lectures and demonstrations that everything was fine. This service "contributed to reducing tension and creating a favorable environment among workers. The results of the measurements proved that the surrounding area was clean, so people lived and worked quietly at the [New] Mine, as well as gathering mushrooms, berries, and fishing in nearby bodies of water in the summer."[14] Yet, as Bellona later reported, "The explosions, however, proved to be of little use: The extracted ore still remains at the site because the area still has no roads to transport it to a processing plant. The government's plan simply had not provided for road construction before the decision to conduct the nuclear blasts was made."[15] One would have thought that in the glorious socialist economy with the emphasis on rational planning they would at least have built roads.

Let the Atom Be a Worker, Not a Soldier

The Brezhnev era has been characterized in standard histories of the USSR as a time of economic stagnation and empty appeals to "trust in cadres," meaning that the nation muddled through the growing challenges it faced in public health, environment, and in declining economic performance, because of the party's inability to embrace meaningful and significant reforms, and because of increasing reliance on fossil fuels to support economic growth. This view certainly holds for the nuclear world. An increasingly complacent and flush Minsredmash found its requests from the government for support for new programs and reactors met with little questioning. Industrial prototype reactors developed in the 1960s included the RBMK under the new head of the Kurchatov Institute, Anatolii Aleksandrov. Aleksandrov also pushed the deployment of a fleet of nuclear icebreakers and serial production of nuclear submarines at Sevmash in the ZATO Severodvinsk.

But the military atom was of far greater importance to the Brezhnev leadership to place nuclear warheads on ICBMs, bombers and submarines. Most important for the Brezhnevites was parity with the United States in the arms race, and the achievement of Détente. During Détente, roughly from 1967 to 1979, Cold War tensions between the United States and USSR eased somewhat as the two superpowers pursued increased trade, scientific, and cultural exchanges, and cooperation with the signing of the Strategic Arms Limitation Talks (SALT) treaties. But Détente was acceptance of international tensions and rivalries, the division of the world into spheres of influence, and approval by the United States of Brezhnev's adventurous foreign policy—until 1979 and the Soviet invasion of Afghanistan. Indeed, the world's nuclear arsenals grew from slightly more than 3,000 weapons in 1955 to over 37,000 weapons by 1965 (United States 31,000 and the Soviet Union 6,000), to 47,000 by 1975 (United States 27,000 and Soviet Union 20,000), and over 60,000 in the late 1980s (United States 23,000 and the Soviet Union 39,000).

It is not the place of this book to discuss the complexities of nuclear doctrine as it evolved from the 1950s when there were a few dozen weapons to tens of thousands of them. Nor is it possible to consider except in passing such concepts as "containment," by which the United States would use its early advantage in nuclear warheads to stop the spread of communism, even to the threat of using them to halt the North's advance with Chinese support into the South during the Korean War in the early 1950s, or MAD, meaning that the two main belligerents, the United States and the USSR, would maintain such large arsenals of nuclear weapons as to render a first strike attempt at "victory" over the other meaningless—and with worldwide devastation as a result. Yet MAD, in all its madness, was considered a source of strategic stability during the Cold War.

The Brookings "Atomic Audit" project (1998) determined that from 1940 through 1996, the US spent nearly $5.5 trillion on nuclear weapons and weapons-related programs, in constant 1996 dollars.[16] Because of secrecy, different ways of counting and so on, it is difficult to determine Soviet nuclear war expenditures. But even if they were only a third or a quarter as much as the United States to build and deploy weapons to achieve parity the amount would be around $2 trillion. And thirty years after the end of the Cold War the United States still spent $37.4 billion on nuclear weapons in 2020 alone, with China spending $10.1 billion and Russia $8 billion.[17]

Consider the nature of the nuclear inventory of the USSR to indicate just how many resources—capital, labor, uranium, social, environmental, political—were devoted to the nuclear enterprise in the name of peace and parity. Toward these end, the USSR developed a massive arsenal of nuclear weapons capable of being delivered on US targets—cities, industries, citizens—by ICBMs, SLBMs, and jet bombers. Norris and Kristensen calculated that "since 1960, the Soviet Union/Russia has built at least 5,000 ICBMs of five generations" with warheads carrying "yields ranging from 220 kilotons to 20 megatons. At their peak in the late 1980s, Soviet ICBMs carried some 7,000 warheads." They noted that the cumulative megatonnage of the Soviet ICBM force typically was three to four times larger than that in US forces, peaking at almost 450,000 Hiroshima-sized bombs in the mid-1970s. They write, "As of 2008, Russia has twice as many ICBM warheads as the United States with six times the total yield."[18]

In an attempt to lessen the risk of nuclear war and provide some measure of stability in relations fraught with danger, misunderstanding and fear, the leaders sought at least some lasting arms control. It helped that, since the Geneva conferences, and continuing through other bilateral and other exchange programs and conferences, many of the actors in arms control efforts had come to know and trust each other. One step in the direction of confidence-building measures was the founding in 1972 of the International Institute of Applied Systems Analysis (IIASA) that centered on the effort to establish bridges of scientific cooperation across Cold War tensions in such areas as AI, climate change and water pollution.[19] Through IIASA and other government and non-governmental organizations scientists, officials, and arms control negotiators achieved some success in slowing the pace of the arms race and in limiting certain kinds of weapons.

Treaties represented hopeful moments, but the manufacture of WMD continued. Two noteworthy treaties were SALT I and SALT II. Richard Nixon (1968–74) sought arms control with the USSR. In May 1972, Nixon was the first US president to visit Moscow. On arrival, Nixon and the chairman of the Supreme Soviet, Nikolai Podgorny, signed an agreement establishing cooperation in environmental protection, a move that gave the Soviets cover from a dreadful environmental record. Nixon and Prime Minister Alexei Kosygin signed another agreement that led to the Apollo-Soiuz space project.[20] The most important event was the signing of SALT I that froze the number of strategic ballistic missile launchers at existing levels, and the Anti-Ballistic Missile Treaty (ABM) that restricted the two nations to two sites for ABMs with 100 missiles each.[21] In limiting defensive systems, the ABM would reduce the "need to build more or new offensive weapons to overcome any defense that the other might deploy."[22] Under SALT I the construction of submarine-launched ballistic missiles on all nuclear submarines was frozen at current levels. Yet the agreement did not freeze the number of warheads which can be fitted to a missile. The United States was more capable in designing smaller devices and hence American missiles with multiple warheads offset the advantage the Russians had with the larger number of ICBMs. Further, under SALT I any further construction of submarine-launched ballistic missiles had to be accompanied by the dismantling of an equal number of older, land-based ICBMs or older submarine launchers.

As reported in *The Guardian*, "At the signing ceremony, there was genial hand-shaking all round, at times excitedly, one Soviet journalist declaring: 'This is not only the first visit of an American President to Moscow—this is a turning point in the history of the world.'"[23] But SALT I did not stop the arms race; the belligerents continued building bombs, debated ways to hide their missiles from detection or harden silos to protect them from nuclear attack, and they fought over whether ABM plans should be completed, how to verify the treaty

Figure 10 Leonid Brezhnev and Richard Nixon conversing at June 1973, Washington, DC, Summit that marked the high point of Détente. Courtesy US Department of State. © Wikimedia Commons (public domain).

and US worries that the Soviets continued to pursue SLBMs with over 100 nuclear submarines at the peak.

Ultimately, of course, as fitting the arms race, the USSR sought to deploy SLBMs through serially produced nuclear submarines that were built at Sevmash in Severodvinsk. The goal was to be able to carry missiles into the world's oceans. A challenge for the Soviets was that most of their SLBMs used liquid rocket fuel with highly volatile and toxic nitrogen tetraoxide and dimethylhydrazine. Accidents involving depressurization of fuel tanks, and poor handling, transport and disposal were encountered, and submarines themselves have been lost or damaged because of reactor leaks, poor damage control, and, especially early on, low-quality control during construction.[24]

Arms control progress was in any event grudging. At the November 1974 Vladivostok Summit, President Gerald Ford and Brezhnev agreed on the basic framework of a SALT II agreement. It included a 2,400 limit on strategic nuclear delivery vehicles (ICBMs, SLBMs, and heavy bombers) for each side; a 1,320 limit on MIRV-ed systems; a ban on new land-based ICBM launchers; and limits on deployment of new types of strategic offensive arms. But many thorny issues remained among the warhead counters: the number of strategic bombers and the total number of warheads in each nation's arsenal. For example, US negotiators believed Soviet Backfire bomber could reach the United States but the Soviets refused to include it in the SALT negotiations, while "the Soviets attempted unsuccessfully to limit American deployment of Air-Launched Cruise Missiles (ALCMs)."[25] Verification also divided the two nations, but "eventually they agreed on using National Technical Means (NTM), including the collection of electronic signals known as telemetry and the use of photo-reconnaissance satellites."[26]

On June 17, 1979, President Jimmy Carter signed the START II treaty with Brezhnev in Vienna, Austria. The treaty limited the total of both nations' nuclear forces to 2,250 delivery vehicles, and there were other restrictions on strategic nuclear forces, including MIRVs. But SALT II never went into effect. The treaty codified parity of sorts. Yet it "did little or nothing to stop, or even substantially slow down, the arms race," and it "met with unrelenting criticism in the United States. The treaty was denounced as a 'sellout' to the Soviets, one that would leave America virtually defenseless against a whole range of new weapons not mentioned in the agreement."[27] Indeed, "a broad coalition of Republicans and conservative Democrats grew increasingly skeptical of the Soviet Union's crackdown on internal dissent, its increasingly interventionist foreign policies, and the verification process delineated in the Treaty," especially after the Soviet invasion of Afghanistan on December 25, 1979. On January 2, 1980, Carter asked the Senate not to consider SALT II for its advice and consent, and it was never ratified. Both Washington and Moscow subsequently pledged to adhere to the agreement's terms despite its failure to enter into force. Carter's successor, Ronald Reagan, a rabid anti-communist and vehement critic of SALT II during the 1980 presidential campaign, agreed to abide by SALT II until its expiration on December 31, 1985, while he pursued the Strategic Arms Reduction Treaty (START). He and his advisors argued that research into the Strategic Defense Initiative (SDI, also known as "Star Wars") adhered to the 1972 ABM Treaty,[28] a position the Soviets fully rejected.

For his part, Brezhnev criticized the United States for a failure to work toward arms control results. On May 19, 1982, he said,

It is no longer sufficient just to speak of peace. Concrete and practical deeds are necessary. A key task today in this respect is to reduce the nuclear confrontation in Europe, a nuclear confrontation that has reached dangerous limits, and to cease further buildup of the nuclear potential. There is a need to prevent a world nuclear conflagration that is in real danger of breaking out at any moment in Europe, where two world wars have already started. The USSR and the United States are soon to resume negotiations in Geneva on limiting nuclear arms in Europe. We shall see how the Americans will conduct themselves—whether they will continue to take their time, while preparing for the deployment of missiles, or will show a desire to reach agreement. The Soviet proposals on this problem are known. We have come out for the total elimination of all medium-range nuclear weapons in Europe. The West contends that this would mean going too far. We have suggested reducing the weapons by more than two-thirds. We are being told that this is too little. Well, let us look for mutually acceptable amounts— we are also ready for larger cuts—on a mutual basis, of course … To facilitate matters, the Soviet Union recently discontinued unilaterally the further deployment of medium-range missiles in the European U.S.S.R. and decided to reduce them by a certain number. I can report that we are already affecting the reduction of such missiles by a considerable number.[29]

Efforts to control nuclear weapons continued under Ronald Reagan and Mikhail Gorbachev who, at the Reykjavik (October 1986) Summit, nearly achieved a bilateral agreement on arms control. They collapsed because the United States would not give up its "Star Wars" anti-missile defense. Still, in December 1987 in Washington the two signed a treaty limiting short-range and intermediate range ballistic missiles (INF) that was important to European allies as well. They feared being destroyed in nuclear volleys that occurred with a conventional force battle between NATO and the Warsaw Treaty countries that began with tanks.

A number of analysts argue that Reagan won the Cold War by refusing to buckle under to the USSR and through massive increases in the US military budget making Gorbachev sue for peace. But the crucial factor was the change in foreign policy—"new thinking"—that Gorbachev advanced. As for Khrushchev and "peaceful coexistence" that abandoned the confrontational policies of Stalin, so Gorbachev recognized that the USSR must work with the other nations of the world and must end the war in Afghanistan. British Prime Minister Margaret Thatcher recognized this when she said the new Soviet leader was someone with whom she could work. Gorbachev was recognized for his determination to wage peace with the award of the Nobel Peace Prize in 1990. To some observers this was an insult to the courage and moral virtues of Andrei Sakharov who won the Nobel Peace Prize in 1975 for standing up to the Brezhnev regime and insisting that the government acknowledge the human rights of its citizens.

Figure 11 Mikhail Gorbachev and Ronald Reagan Sign INF Treaty, December 1987, Washington, DC. © Courtesy of Ronald Reagan Presidential Library (public domain).

Brezhnev's Bold Nuclear World

The Cold War nuclear industry expanded rapidly from 1945 until the Chernobyl disaster of 1986. Nuclear weapons production took first position, while civilian applications were far less important, although an unsuccessful campaign to build literally dozens of 1,000 MWe power generating reactors based in part on their serial production in the Volga River Atommash factory unfolded in the 1970s. Both the civilian and military nuclear industries stand out for their contribution to significant hazardous waste problems. They include the jettisoning of submarine reactor vessels in the Arctic and Pacific Oceans and the haphazard disposal of solid and liquid waste including in sites that have not been recorded officially. They extend from naval bases in Murmansk to the Far East, and from Novaia Zemlia, site of extensive nuclear testing in the Far North to Kazakhstan to the south central part of the empire.

VVERs, RBMKs, and breeders. Nuclear warheads. Building dams and mines with PNEs. Nuclear-powered satellites. By the 1970s the nuclear enterprise had become one of the most respected—and well-funded—branches of Soviet industry. Its military capabilities had achieved parity with the United States, with tens of thousands of warheads fitted onto rockets, bombers, and submarines. Its reactors moved icebreakers across the Arctic. Reactor "parks" were being planned near major cities including several proposed to provide steam heat for industry and homes. If many of its operations—both peaceful and nuclear—were secret to the public and carried out in closed institutes or ZATOs, then still the average citizen welcomed the efforts of MinSredMash for keeping them safe, warm and well-illuminated.

CHAPTER 5
NUCLEAR DISINTEGRATION

On April 26, 1986, the fourth reactor of the Chernobyl Nuclear Power Station (ChNPP) exploded, releasing large amounts of radioactivity into the environment. The accident happened during the cooling down of unit four before scheduled maintenance. Operators intended to permit the turbines to spin from their own momentum after the shutdown to see how long they would continue to generate electricity. In the middle of the shutdown, the Kyiv grid called for more electricity. But rather than bring the reactor up to full power, a timely and costly process which would prevent the experiment, the operators disabled various safety systems and removed control rods from the reactor core in order to keep electricity production up—but without any safety margin. Already at low power, and without safety or control systems in place, a so-called positive void coefficient came into play. In this case the Chernobyl reactor experienced an exponential surge in power, the reactor core overheated, the cooling water boiled out of the core (increasing the power further), the core melted down, and a chemical reaction of steam with metal and/or graphite yielded an explosive mixture of hydrogen and oxygen. Two powerful explosions ripped through the reactor destroying it and lifting its lid—at 2,000 tons—into the air and down on its side, and destroying the roof of the standard factory building. One hundred to two hundred MCi (megacuries) of radioactive substances filled the environment over the next ten days falling onto the land and entering the water around the station, and also into the atmosphere where it spread through the northern hemisphere, hitting Ukraine and Belarus especially hard. Fuel rods, burning graphite, and other material scattered on the ground and the roof of reactor unit three next door, which, against the regulations, had a flammable bitumen cover, and instantly caught fire. Inside several other areas caught on fire, but through the heroic—and mortal—action of the firefighters, the most dangerous fire spots were extinguished by 5 a.m. But the core of the uncontrolled nuclear reactor was open, and its graphite burned, emitting visible fumes and invisible radiation into the environment. The base of the reactor was forced down four meters, the explosion having demolished the supporting structure. Highly radioactive lava of the melted nuclear fuel and construction/building materials flooded lower corridors and rooms of the building.[1]

The authorities ordered the evacuation of the 45,000 residents (including 17,000 children) of Pripiat only thirty-six hours after the explosion. Evacuation of such heavily contaminated settlements as Chernobyl town (with 20,000 inhabitants) and the Gomel region of Belarus followed later in May. But in all there was an unforgivable seventeen-day delay before the Soviet government informed the Soviet people with detailed information about the Chernobyl disaster. Sometime in August, the evacuation of 166,000 people from eighty-eight towns and villages in Ukraine, Belarus, and Russia was complete. In addition, 60,000 cattle were transported from the zone, some of which made it into sausage sold throughout the USSR—except, by secret order, in Moscow province where the elite lived. The authorities used 8,500 vehicles including 2,500 buses in the evacuation—many of which were later buried in a "graveyard" near Chernobyl. Eventually, they established a thirty-kilometer diameter exclusion zone surrounded by barbed wire and protected by armed guards. Soldiers were sent in to shoot all animals including pets lest they escape the zone.

A lasting monument to "nuclear Russia" is the Shelter Object (in Ukrainian, Ob"ekt "Ukrittia") that was built in several months by poorly equipped workers to cover the open reactor as quickly as possible to limit radioactive contamination from spreading further. Facing a dangerous task, the authorities requisitioned soldiers, miners, concrete pourers, and others from around the nation to the site. As such, the Sarcophagus, as it became known, is a symbol both of technological failure and of the power of the state to command workers to toil in danger. But in some ways the authorities had no choice. Within the reactor remain 200 tons of the melted, highly radioactive core, 30 tons of dust, and 16 tons of uranium and plutonium—all lethal to humans after short term exposure. Just three weeks after the explosion, engineers of MinSredMash began design of the sarcophagus which commenced in June and was completed in November. Working in shifts fifteen days on, fifteen days off, the men used 400,000 m³ of concrete and 7,300 tons of metal. Perhaps 200,000 men were involved in all.[2]

At first the authorities were convinced they could continue to operate the three other units safely on site. On October 1, 1986, unit 1 began operation again, and on November 5, unit 2, while on October 2 the government ordered the construction of the town Slavutych to support power operation. In essence, during the first months and years after the disaster the authorities focused their efforts on what they called "liquidation" of the accident's consequences that would allow the quick return to "normal," not only of the undamaged units of the ChNPP, but of the landscape, too. The men (they were mostly men) removed contaminated soil, flora, and fauna and buried it in the zone. They then buried or isolated the equipment—buses, bulldozers, trucks, helicopters, automobiles, and so on used in calming the reactor after the explosion and then in liquidation itself. The liquidators all received some doses of radiation, some of them very high doses of radiation and developed cancers and immunodeficiencies. It seems that we will never know how many people paid the consequences of their heroic work with bad health, cancers, or death because of poor record keeping and poor access to health care.[3]

Chernobyl as a Symbol of Soviet Radioactive Failure

Many people gained familiarity with the Chernobyl disaster through the superb HBO series "Chernobyl" (2018). A huge and impressive literature addresses the major aspects of the disaster: its prehistory and planning, glorious construction, the goals of building a "park" with up to ten massive 1,000-MWe RBMK units, the construction of Pripiat, the city that arose but three kilometers from the station to support reactor operation, the fateful "experimental" shutdown that doomed the reactor and revealed the lack of safety culture; the dangers of nuclear power generally, and indeed the unstable design of the reactor; the first few days of confusion, delay, lies, and obfuscation, and tardy evacuation of tens of thousands—and later hundreds of thousands of affected individuals who lived and continue to live in areas of dangerously high radioactivity; the heroic, if mad cleanup by "liquidators"; the building of a "sarcophagus" to entomb the dangerous reactor guts, and the construction of another entombment in 2017 to protect the sarcophagus from a disastrous collapse; the race to get reactor units on line again within six months, requiring that workers continue to toil in high radioactivity; and the building of the city of Slavutych 50 km from the reactor by 1988 to house the workers of the Chernobyl station that continued to operate until 1999, itself both a monument to Soviet planning and determination in that the beautiful city was built with a short time horizon, and to the incompetence and inhumanity of central planners and party officials, for which of them would live and work in Chernobyl region after the devastating accident in 1986?[4]

Chernobyl was supposed to be the greatest achievement of the Soviet nuclear establishment with ten massive 1,000 MW reactors when complete. On the eve of the world's most famous technogenic accident, there were four operating reactors, and two others in an advanced stage of construction. These reactors were built near the Pripiat Nature Reserve on the Pripiat River that flows into the Dnieper River and directly to the capital of Ukraine, Kyiv, only 90 kilometers away, as a suggestion that the nuclear industry could build powerful machines in an Edenic environment, a kind of reactor in a garden—or "park" as engineers and journalists referred to this massive nuclear wonderland. And it was a park. A town of 50,000 people, Pripiat, was built to house the workers of the ChNPP, who spent the weekends at playgrounds with their children, playing soccer, fishing in nearby rivers and streams, gathering bountiful berries and mushrooms in the woods. Geese came to winter at the Venice-like canals built to carry cooling effluent from the reactors.

Construction at ChNPP commenced in 1970, the first reactor was connected to the electric grid in 1977, the second unit in 1979, the third in 1981, and the fourth in 1983, all RBMKs as opposed to the safer PWRs with containment vessels. At other Ukrainian nuclear sites, the authorities opted for VVERs that were brought into operation during the early 1980s: two at Zaporozhskaia power station, two at Rovenskaia power station, and two at the South Ukraine power station. Thus, by the time of the Chernobyl accident ten reactors were operating on the Ukrainian territory (and in 2022 fifteen reactors are on line in Ukraine that provide over 50 percent of the nation's electricity).

On the eve of the breakup of the USSR the scientific establishment remained all-powerful. The country claimed one-quarter of the world's physicists and one-third of its engineers who were held in high esteem by officials and citizens alike. Even if there was criticism for the failure to introduce achievements from research and development into the stagnating economy, most individuals blamed the planning system and bureaucrats, not scientists. The space and nuclear establishments seemed beyond reproach with the latter gearing up to double or treble nuclear generation capacity by the end of the century. Increasingly, in the other republics, and also in the socialist nations of Eastern Europe, nuclear training programs, research facilities, and power-generating reactors were being built and expanded, to the delight of local elites—and citizens—who relished being part of nuclear modernity.

The disaster shattered the complacency of nuclear engineers and citizens alike about the apparent safety of living in a modern industrial society. After a seventeen-day delay in admitting the extent of the disaster to the nation's citizens, Gorbachev finally addressed the nation. Yet the accident had far more than significant and ongoing, if officially downplayed, radiological impacts. Chernobyl required the Communist Party leadership precisely to confront glasnost over systemic failures in economy and society, a process that angry citizens ensured would carry beyond the dangerous, if heroic effort to tame the smoldering reactor to industry and agriculture at the ends of the empire. Chernobyl's place in the history of the last years of the USSR was to accelerate the collapse of the USSR. Akin to radioactive decay of the uranium atom, by the late 1980s the Soviet Union had begun to spin apart. The economy had proven incapable of innovation and unreceptive to reform. An increasingly urban and well-educated population had come to understand there were few rewards in society and that they lived a lie of "developed socialism." Demographically, non-Slavs were the largest population group in the empire, and manpower and loyalty concerns came to occupy Soviet leaders. If most citizens still feared the United States and an attack by the west, then there was little else they could see as a benefit bestowed by an increasingly corrupt and tottering regime. The stores were often bare, the quality of apartments minimal, and the nature of labor and work was unfulfilling. People lived according to the sly joke presented in the form of a Soviet slogan: "They pretend to pay us and we pretend to work."

Gorbachev, Chernobyl, and the Need for Reform

Mikhail Gorbachev pursued perestroika (reform or revolution) through glasnost (openness or transparency) to expose past Soviet failings and encourage public participation in the effort to rebuild economy and society. Gorbachev intended to modernize the economy, encourage innovative business and production practices, weaken entrenched bureaucrats, all with the involvement of citizens to expose past and potential practices of corruption, toadyism, and laziness. Yet Chernobyl accelerated the breakup of the USSR as directly as did Mikhail Gorbachev's efforts to reform the Communist Party and state apparatuses. First, Gorbachev came to understand that the disaster revealed just how much in need of reform the system had become. Rather than respond quickly and properly to the disaster, at each level of authority officials spread lies, obfuscated the true extent of the crisis, and slowed the proper response. And ultimately, rather than come to grips with the fact that the industry had built an inherently unstable reactor and the system that produced it, the authorities found human scapegoats to punish while the public remained dangerously in the dark about nuclear technology.

Gorbachev himself was at the top of this decrepit system. He failed to push for the facts from Chernobyl immediately after the accident, but for two weeks was paralyzed by the usual need to cover up or lie about the accident. He accepted the words of officials on site. Only later, after journalists visited the site, did he realize the extent of the disaster and the failure of the entire civil defense system to provide for those affected. Medical and fire departments were poorly equipped, there was no protective clothing, nor adequate numbers of radiation sensors or Geiger counters censors, no stores of iodine pills, and so on. There was only chaos. Gorbachev realized that the Soviet culture of secrecy was at fault. On top of this, he soon came to understand enormous economic costs of handling the accident. At a July 3, 1986, politburo meeting, he himself exploded at the lies of the officials and scientists of Soviet ministries and research centers. After Chernobyl, Gorbachev concluded that he could never accept a military budget or arms control proposal based on the premise that a nuclear war might actually be fought. One of the main purposes of *glasnost*—itself triggered by Chernobyl—was to break the military's monopoly on national security information.[5]

Nuclear Wastelands[6]

By the eve of the breakup of the USSR, the nation was dealing not only with the fallout from Chernobyl, and with the need to provide stewardship to an increasingly unstable stockpile of nuclear weapons and nuclear submarines, but with the vast quantities of solid and liquid radioactive waste produced in fulfillment of military and civilian programs. The USSR—Russia—has more waste than any other country in the world, and only in the 2000s and 2010s began to inventory it and attempt to store it safely. Researchers determined that they do not know the location of all of the waste, nor necessarily of its disposition, quantity, and state when they know of its location. The effort to deal with the waste more carefully began precisely in the Gorbachev era when exposés of past accidents, haphazard disposal, and cavalier treatment of people and nature were published in newspapers and journals and screened on TV. People learned chapter and verse of the environmental and public health costs of the Soviet development model, and perhaps nowhere more than in the nuclear sphere was the litany of disasters and malfeasance in management and disposal greater. In post-Soviet Russia in the Yeltsin years the government made more of an effort to address nuclear waste problems, at least so far as to catalog the situation. But within state security and military organizations that had great continuity and personnel with the Soviet era an attempt to silence whistleblowers persisted. Two accidents dating to the dawn of the Soviet nuclear age, that came to light only in the 1980s, indicate the extent to which the authorities used Cold War secrecy to deny the public information about the risks they faced from the nuclear enterprise, and the fact that scientists maintained their belief they could manage the waste properly, or find technological solutions to accidents whenever they occurred.

In the heat and dangers of the Second World War and then the Cold War, Soviet strategists and planners located armaments industries in the Ural Mountains. Having established tank, airplane, and metallurgical plants, they logically determined to build nuclear weapons facilities in the region as well to take advantage of an existing military-industrial labor force and the distances from the nation's borders. As noted earlier, they also employed tens of thousands of prisoners in building the nuclear enterprise. Taking the same attitude toward the natural environmental as they did toward workers, they picked sites on rivers and lakes to ensure access to copious amounts of water for cooling and other industrial purposes, and they eventually used those waters to dilute and dispose of the toxic chemical and radioactive wastes produced in weapons manufacture. The result was to create some of the most polluting industries in the world in previously quiet farmland and along relatively pristine rivers and ponds.

One of the central sites of nuclear violence was the Techa River. The Techa River basin had been the home of farmers, fishermen, trappers, and their families who lived a quiet life until the opening of the Maiak plutonium production facility in 1948 with a series of production reactors coming on line by 1955. Prior to opening of

Maiak, the Techa River flowed out of its source at Lake Irtyash and then through Lake Kyzyltash. A number of other tributaries, creeks, brooks, and small rivers enter the Techa, later as the right tributary of the Iset' River, on its way to the Tobol River. But the river's life changed with Maiak. In the late 1940s Maiak managers began storing high-level waste in tanks that were inadequate to the task given the great volumes of liquid waste of low- and medium-level wastes. They could not decrease the waste radioactivity or the concentration of radionuclides to less than 10 million Ci/cm^3, and therefore dumped the waste into the Techa River. Scientists apparently believed that the swampy and slow-moving Techa was an ideal place for sedimentation of radioactivity, even if the tributaries of the Techa, including little streams and brooks, dried up in the summer, and even if there were villages and farms along the river. From 1949 to 1956 Maiak dumped an estimated 76 million m^3 of radioactive waste water into the Techa. In a word, the lakes and river became a plutonium production and waste disposal site.[7]

The area was a site of nuclear waste disasters, both chronic and catastrophic. On September 29, 1957, twenty MCi (740 PBq) of radionuclides were released by a chemical explosion in a radioactive waste storage tank in Kyshtym, 10 percent of which spread beyond the site to form the East Urals Radioactive Trace (EURT), with tremendous long-term health impacts, as yet poorly understood. The waste spread over 20,000 km^2 where more than 270,000 people lived. Over the next days and weeks officials arrived and required peasants to destroy livestock, they leveled over twenty villages, and evacuated over 11,000 people. Battalions of prisoners were brought in to use wheelbarrows and shovels to turn the soil and prevent the further spread of the poison. Residents knew that something had occurred, but what precisely they never understood.

Natural disasters contributed to the spread of radioactivity along the river's flood plain. In 1951 radioactive water polluted the bottomland of the Techa because of such a flood; inhabitants used this area as meadows. For inhabitants of the Techa shores, the river was a source of food—and hence irradiation, both external radiation from gamma rays near the river and internal radiation form radioactive isotopes that entered into organisms through water and food. The largest doses (50 to 100 centiSv/year) occurred among the people of Metlino village, and many of them got ill. Internal radiation of elderly people mounted to about 4.6 milliCi from the river. The Ministry of Public Health refused all requests to support the population or recognize their injuries.[8]

Eventually recognizing that something must be done, the authorities ordered the building of a series of dams and holding ponds to cool the waste and with the prediction it would settle into the muck and not migrate further. While the system of ponds, used as sumps, and canals slowed the spread of radioactivity, it did not prevent it. In 1954 the authorities decided to build another reservoir with volume of 29 million m^3 in which radioactive water would accumulate and evaporate naturally. But the volume was inadequate since the plant dumped more than 40,000 m^3 daily, or 15 million m^3 annually. By autumn 1959 the reservoir was full. This realization led to the construction of other dams/reservoirs between 1956 and 1963. Once again, this slowed the pollution of bottomland, but did not prevent radiation spread. Ultimately, the authorities were forced to evacuate some 7,500 people from nineteen villages from 1955 to 1960, too late to protect them.[9]

Similarly, reckless behavior, about which the public learned only in the Gorbachev era, led to the infamous Lake Karachai disaster in 1967. The cascade of reservoirs 3, 4, 10, 11 just below Lake Kyzyltash (reservoir 2) with a total surface area of 84 km^2 and volume of 380 m^3 was intended to isolate 90 percent of the radioactivity of strontium and cesium, but likely achieved a level only of 80 percent. To lower the waste level, in September 1951, the Soviets stopped discharging the diluted high-level radioactive waste directly into the Techa and instead diverted it into Lake Karachai. The 1967 accident occurred when a hot summer followed a cold winter, the water evaporated, and the lethal radioactive dust from the lake bed blew over a vast area, about 1,800–2,700 km^2 including the reactor site and 41,500 people in sixty-three villages, some of whom had already suffered from Kyshtym. Only in the 1990s did the authorities fill the lake with hollow concrete blocks, then rock and soil to keep the sediment down, cover the lake, then pump and treat the water. By then, the lake had accumulated 120 MCi of long-lived radionuclides Cs-137 and Sr-90. Visiting delegations in 1990 detected high levels of radiation even "a few hundred feet" from the lake, higher where the radioactive effluent was discharged.[10]

Radioactive risks were not limited to far away farmland, but were built close to the Kremlin. Putting secrecy and defense concerns first, even as Moscow encircled October Field, KIAE built a series of nuclear reactors and maintained the spent fuel (SNF) and other waste on site. Only in 2019 did the institute reach agreement to transfer the SNF that had accumulated for seventy years to Maiak. In all, there were some 900 spent fuel assemblies weighing about six tons and total waste (not including contaminated soil) of around 1,200 cubic meters at 2,000 tons.[11]

Disintegration

As news of accidents and disasters, past and present, spread cross the USSR, demonstrations broke out. In Bashkirostan construction of two PWRs was frozen in 1990 (although the Russian government has begun pressing to open the project again).[12] In Lithuania, a major impetus to the independence movement was the presence of the Ignalina NPP with plans to build perhaps a third Chernobyl-type reactor; the Saujudis national liberation movement, founded in June 1988, built on worries of another Chernobyl to push for true independence from the USSR.[13] In Ukraine glasnost and revelations about the true scale of the Chernobyl disaster fueled anti-nuclear, environmental, and nationalist movements. Greens were represented by the organization Zelenyi Svit created in 1987 that came to life as an organization of Ukrainian writers preoccupied by the environmental degradation of nation and the consequences of the Chernobyl disaster. The writers were soon joined by the scientists and other representatives of the Ukrainian intelligentsia. Yuri Shcherback, a Ukrainian writer and a medical doctor who published a novel on the Chernobyl disaster in 1988, became the chairman of the Zelenyi Svit.[14] Ukrainian nationalists were represented by Ukrainian Popular

Movement (Rukh), established in early 1989. The two movements were closely linked in the period between 1989 and 1991, with Rukh members strongly supporting anti-nuclear claims and protest actions.[15]

The policies of the Soviet authorities to secure the fastest possible return to normal of the Ukrainian as well as Belarusian people affected by Chernobyl could be defended only because of the official cover-up of the true scale of the disaster. Thus, during almost three years after the disaster all the media controlled by state and the Communist Party were filled with accounts of the heroic effort of the soldiers and firemen that succeeded in reestablishing control over the reactor and cleaning up adjacent territories from radioactivity. These accounts were complemented by the reassuring and optimistic statements about the dropping levels of the radioactivity and close to normal conditions for everyday life for local people. The authorities even moved ahead with plans to continue construction of six more reactors at the ChNPP.

Protest broke out. Both nationalists and environmental activists saw this as representative of colonial power: officials in Moscow made decisions about building NPPs in the republic without consideration of the dangers and risks for Ukraine. Further, Ukraine did not have its own branch of Minsredmash, nor its own regulatory agency. This colonial attitude of the Moscow authorities fueled resentment. In 1989 and 1990 anti-nuclear activists gathered petitions, organized strikes, pickets, and blockades against the construction of the new units at the Kmelnitsky, Chigirin, Crimean, South Ukraine and Chernobyl NPPs. The anti-nuclear mobilization led the Ukrainian Parliament (Verkhovna Rada) in August 1990 to declare a moratorium on the construction and commissioning of new nuclear power units. (By the 2000s Ukraine had embarked on the difficult path to achieve energy independence in part through nuclear power.)

By 1988 environmental and anti-nuclear movements had broken out across the nation, from Ukraine to Lithuania, and to autonomous republics, with some activists seizing on the claim that Moscow had exploited the periphery with dangerous extractive industries and unsound technologies. (Lithuania abandoned its two RBMKs in the

Figure 12 Cows near a radiation sign on the Techa River in the Ural Mountains, the site of severe accidents and extensive radiation pollution. © Wikimedia Commons (public domain).

2000s as a requirement to enter the EU[16]). The USSR disintegrated five years after Chernobyl on December 25, 1991.

Ultimately, the legacy of Chernobyl was this disintegration of the USSR and the collapse of one of two great Cold War nations into the muck of political corruption, economic decay, and failed big projects. Both Ukraine and Belarus in June 1990 declared their sovereignty as non-nuclear states. But many symbols of technogenic failure fill the landscape as reminders of their participation in the Soviet socialist experiment. The first was the New Safe Confinement to entomb the Chernobyl sarcophagus itself. The sarcophagus, built hurriedly in dreadful conditions was in danger of collapse, and almost a sieve during rain storms. The New Safe Confinement is designed to prevent the release of radioactive contaminants and eventually to facilitate the disassembly and complete decommissioning of the reactor. Built at some distance away, then moved pneumatically along train tracks to cover the sarcophagus, it was discussed among the G7 nations, Ukraine, and Russia from the late 1990s but only finished in 2017.

The 110-meter tall and 257-meter wide arched structure was late in construction owing to finance problems and the challenges of building a new foundation in radioactive conditions.[17]

Two other memorials of disintegration are the abandoned, half-finished shells of nuclear power stations, abandoned in the fire of protest of the late 1980s, an abandonment confirmed with the breakup of the USSR. This was fortuitous in many cases, for example, the Crimean NPP, to be built in a seismically active region. The abandoned NPP served from 1993 until 1999 as the stage for the annual "Republic of Kazantip" musical festival in July and August; one of the inspirations of the Orange Revolution in Ukraine in November 2004–January 2005, that led to new elections, Nikita Marshunok, was "president" of the republic.[18] The second is the 2,000-ton lid to Chernobyl 4, itself blown into the air by the explosion, and now resting nearly vertically, a vivid reminder of the recklessness of building a reactor without a containment vessel.

And in the exclusion zone bountiful berries, delectable mushrooms, succulent fish, and wild boar, all drenched in radioactivity, still serve as food to local people who returned to zone or live nearby in farmland and forest in Belarus, Russia, and Ukraine.

CHAPTER 6
NUCLEAR RENAISSANCE

Construction on nuclear power stations in Russia ceased during the prolonged economic crisis of the 1990s. Exposés and official studies promoted by the Yeltsin administration revealed the true extent of radioactive pollution in all regions of the nation and especially in Arctic waters and the Urals region. The Chernobyl cleanup proved far more costly than envisaged, especially in Ukraine and Belarus, and the hastily built sarcophagus required a "New Safe Confinement" that slid into place only in 2017. Atommash was bankrupted in the 1990s. In summer 2001 President Vladimir Putin signed a bill to permit Russia to import spent nuclear fuel from abroad to earn $30 billion in hard currency for the nuclear industry, if promising to use it for environmental cleanup.

But leaders of the Russian nuclear establishment never lost enthusiasm for new projects, and with the economic recovery of Russia flowing forward on barrels of oil they confidently forecasted rapid and sustained growth of nuclear generating capacity into the twenty-first century. Specialists at the Kurchatov Institute predicted 50 to 60 GWs of installed capacity by 2030, that is, the construction and operation of at least 25 new 1,000 MWe reactors in less than twenty years. Rosatom is now propping up Belarusian dictator Aleksandr Lukashenka with his own 1,000 MW PWR at Astravets, Belarus. It seems that, without dropping a control rod, the nuclear enterprise had maintained its momentum and ambitious visions, and as the resource economy recovered, it drew heavily on the government's petrodollars to proclaim a nuclear "renaissance." Its spokespeople spent lavishly on PR campaigns to convince the public of nuclear power's clean safe future. In 2003, as a sign of a new, friendly, even sexy and friendly nucleus, the industry began an annual "Miss Atom" contest with bikini-clad employees of the various institutes and NPPs of Rosatom competing for first prize. By 2007 officials had created the massive new state corporation, "Rosatom," with capitalization valued at $40 to 50 billion that consists of 350 enterprises and over 200,000 employees.

The renaissance, involving the reincarnation of Minsredmash in Rosatom, has been based on three major programs: the rapid construction of VVERs at home and abroad and the continued operation of 11 RBMKs on Russian soil, the expansion of a long-pursued breeder reactor program, and the design and manufacture of compact, transportable units (floating reactors, third-generation icebreakers to keep Arctic drilling and commercial sites open year round, and proposed nuclear natural gas tankers and nuclear-powered ocean-going oil derricks, the latter based on existing submarine technology). See Figure 13.

Rosatom and its various power generation, icebreaker, and other subsidiaries confidently use the word "renaissance." Currently, it operates thirty-two nuclear power reactors with the overall installed capacity of 24.2 GWs (gigawatts) at ten power stations. The stations account for about 20 percent of domestic electricity generation. The share of nuclear generation in the European part of Russia reaches 30 percent, and in the northwest part of the country—37 percent. The industry comprises enterprises of nuclear fuel cycle, nuclear power engineering, nuclear weapons applications, and research institutes. Rosatom unites a number of enterprises of nuclear power engineering, as well as of nuclear and radiation safety, the nuclear weapons complex, and fundamental research. Nuclear power engineering is a major sector of the Russian economy, and crucial to the state for symbolic and geopolitical reasons.[1]

Rosatom has drawn on continuity with Soviet nuclear programs to inspire today's plans even twenty years after the collapse of the USSR. First, while acknowledging at times the high capital costs per kilowatt hour installed capacity, Russian engineers continue the claim that nuclear power is the only alternative to fossil fuels. If not "too cheap to meter," as the world physics community claimed in the 1950s, then nuclear power will serve modern industrial society well into the twenty-first century when peak oil and gas have long passed, and without emitting noxious greenhouse gases. Second, engineers maintain the conviction that they can adopt standardized designs for reactors as a way to keep costs down, relying ultimately on "serial production" of reactor components, vessels, and plant facilities. The French example provided hope. France produces nearly 80 percent of its electricity from fifty-nine PWRs largely of the same design, and it has an aggressive international sales program, although currently it has been buffeted by significant cost overruns and lengthy delays in reactor construction. Russian officials believe they can avoid cost and delay problems, but achieve the same high level of standardization as the French industry did through Atommash. Finally, as before, today's nuclear engineers believe that they will soon move to a closed fuel cycle using liquid metal fast breeder reactors (LMFBRs). In this belief they must ignore, or downplay, a series of significant problems that have plagued breeders throughout their history.

In all of these ways, hubris and technological momentum are the central aspects of the nuclear industry from its founding in the 1950s. Nuclear power remains a panacea for power production, transportation, and uneven geographic distribution of Russian resources and population. Engineers are certain they can design safe reactors with a variety of different applications. They have the support of the state. Leaders and ministers see nuclear power as a symbol of great power status and support billion-dollar expenditures accordingly. And, as they have for fifty years, program representatives assert they have solved any problems of waste, spent fuel storage, and safety, even as waste remains a fateful consequence of Nuclear Russia.

Nuclear Russia in the Twenty-First Century

Russian planners, engineers, and state leaders have determined to move the nation forward along nuclear rails with an effort to create on closed fuel cycle, where spent fuel is partly reused, combined with enriched uranium or plutonium, and reloaded into reactors, on a foundation of PWRs and breeder reactors. They also intend to keep the submarine builders of the Cold War busy building floating nuclear reactors of various sorts—icebreakers, and oil/gas industry tanker and drilling devices, and of course submarines. With such an array of atomic machines Rosatom and its shipping division, Atomflot, sees itself as a kind of ministry of the Arctic, like Stalin's Main Administration of the Northern Sea Route, Glavsevmorput, capable of powering hydrocarbon development, keeping the Northern Sea Route open, and securing the nation's borders. And Rosatom sees itself as a leading international company, capable of competing with China, Korea, and the United States in sales of PWRs and floating stations abroad to Iran, Finland, Bangladesh, Belarus, and elsewhere.

The next generation VVER1200, the flagship reactor and the core product of Rosatom's program, largely produced at Atommash, features improved performance, additional safety systems, and advanced containment. It has a 20 percent higher power capacity while having a size comparable to VVER1000, an extended sixty-year service that seems unrealistic given the tremendous heat, pressure, and radiation stresses on reactors, and is apparently earthquake proof, its designers claim.[2] As for RBMKs, the Russians believe they can manage still operating units in safe regime, in spite of the lack of containment, rapidly accumulating spent fuel on site, and inherent instability flaws. With license extensions, Kursk units 1–4 and Leningrad units 1–4 will continue to operate into the late 2020s, and Smolensk 1–3 into the 2030s. Units outside of Russia in Lithuania and Ukraine have been permanently closed, in the former case as a prerequisite for joining the EU, in the latter case because of the Chernobyl disaster. In all, this is a program of grandeur based on the institutions and visions of the Soviet past.

Atommash rejuvenated in the 2020s as Russian orders at home and abroad have ticked up noticeably. Iran's Bushehr facility has one operating VVER, with construction on a second, and agreement in principal between Russia and Iran to build two other units. In India the Kudankulam NPP has two VVERs on line with four others to be built. Russia has even engaged talks with Namibian leaders to build a floating NPP to produce electricity and desalinate water, although beyond talks no concrete steps have been taken. Yet even with the

Figure 13 A Soviet postage stamp commemorating the twentieth anniversary of the Atomic Icebreaker "Lenin" (1978), the flagship of the Soviet nuclear icebreaker program. Russia's Atomflot is continuing to build nuclear powered icebreakers in the 2020s. © Wikimedia Commons (public domain).

"renaissance," there remain problems with construction, costs, and delivery with Russian reactors as for French or other reactors. For example, Belarus, the nation most damaged by Chernobyl fallout, has determined to buy two third-generation VVER1200 reactors for its Astravets NPP, only 45 km from Lithuania's capital, Vilnius, at the consternation of many European observers of Russian safety and construction practices. One reactor has come on line and is operating at half-power. But in July 2015, the builders dropped the first 330-ton pressure vessel delivered, damaging it, and requiring its replacement. Rosatom and Minsk officials were silent about the incident up until a whistleblower reported to the media. This accident led Rosatom to insist that there was no damage to the vessel, and it agreed at great cost to replace it only later after first claiming that it had "touched the ground softly."[3]

In the domestic market over the next decades Rosatom intends to build scores of reactors at construction and licensing speeds never before encountered in the international industry, and to increase the share of nuclear electrical energy nationally to 50 percent by 2050.[4] In a series of planning documents since the 1990s the industry has advanced optimistic targets largely based on PWRs, but also breeders and district heating reactors ("AST") with license extensions of existing reactors where financially feasible. By mid-2000s it was clear that the 1998 Program was unrealistic; only three new reactors came on line in 2001, 2004, and 2010 respectively. But Rosatom announced in October 2006 another plan to bring online ten new power units between 2007 and 2015. This bold announcement gave way to yet another forecast of at least thirty reactors to be brought online by 2020 (also *not* realized) including floating nuclear power stations.[5] These documents indicate that the Russian nuclear industry has recovered from its self-inflicted wound of Chernobyl, even if the environmental and public health costs of the Chernobyl disaster will be paid out for some time to come; indeed, Chernobyl is rarely mentioned, or is an afterthought among engineers, although President Putin commemorates the disaster annually on its anniversary.

On top of PWRs and the last RBMKs, Rosatom renaissance includes ratcheting up construction of standard liquid metal fast breeder reactors (LMRBRs). As noted, breeders transmute ^{238}U that is non-fissile into ^{239}Pu that is fissile, and in breeders spent fuel can be reprocessed and used again and again in a reactor, in theory until all the energy in it is extracted, but leaving waste that decays

Figure 14 "President Commands Fourth Block of the Rostov NPP to Come Online by Video Connection." Source: President of Russian Federation, February 1, 2018, at http://kremlin.ru/events/president/news/56767/photos/52356. © Office of the President of the Russian Federation (public domain).

over hundreds of years. Reusing the fuel—closing the fuel cycle—is hardly clean and far more expensive than the open cycle in PWRs, although the breeder's promoters rarely say this out loud. But nuclear Russia intends to be the major player in this technology in the world, although China's first CFR-600 (600 MWe) LMFBR is scheduled to be operational in the mid-2020s—and will produce weapons grade plutonium.

For Russia the breeder journey began in the late 1940s and early 1950s under Aleksandr Leipunsky with a series of experiment units. Hopeful engineers then built the BN350 (MWe) in Kazakhstan on the shores of the Caspian Sea to desalinate water and support a burgeoning petrochemical industry in the desert in the 1970s, then moved along with the BN600 (600 MWe), BN800, and a now planned BN1200.[6] But these reactors have had a checkered history of cost overruns, sodium spills, and fires. Frighteningly, engineers considered building the BN-600 reactor at the Maiak

Chemical Combine, site of the Lake Karachai and other technogenic disasters that would have resulted in an entire region of the nation devoted to plutonium. They settled instead on construction at the picturesque Belioarsk NPP in Zarechnyi the foothills of the Ural Mountains. During the operation of the BN600, there have been at least twelve cases of ruptures, twenty-seven leaks, five of which with radioactive sodium, fourteen with sodium fires, five because of improper repairs, and in one case a leak of 1,000 kg of sodium; sodium conflagrates in contact with water or oxygen.[7] In January 1987 BN600 fuel rods melted. Total radioactivity was about 100,000 Ci, a level four accident according to International Nuclear Event Scale indicating a serious accident with "local consequences." In September 2000 personnel mistakenly cut power to the station; three seconds later the BN600 scrammed, steam was released, and according to an expert commission only several minutes remained before a catastrophic accident would have occurred.[8] Plans to build four BN800s were stopped after Chernobyl, but in 1997 the LMFBR program took off again, and it has wide support in the industry.[9] An expert on many subjects, President Putin claimed that fast reactors are "technically quite feasible," and he acknowledged their importance as confirmation of Russian nuclear leadership. Late in 2016 Russian engineers in Zarechnyi began operation of the BN800. It is no coincidence that Beloiarsk operators pushed to bring the BN-800 on line on December 22, "Day of Energetics," that celebrates Lenin's State Program for Electrification, GOELRO, making explicit the tie between Russia's nuclear program today and its Soviet heritage.[10]

Next on the drawing board is the BN1200 (1200 MW) that is expected to become the mainstay of breeders, with a new design that, engineers claim, will consume 50 percent less steel, with the number of primary loop valves reduced from 500 to 90, and the piping will be 30 percent shorter. If fewer valves and pipes, then lesser risk of leaks? The former head of Rosenergoatom, Romanov, asserted that three BN1200s would be online by 2030 in an unheard-of fifteen-year construction period.[11] Since the mid-2010s, there have been conflicting reports when the BN1200s will actually commence construction and how many will be built.[12]

Floating through the Arctic

Floating nuclear power stations have been the dream of Russian scientists and engineers since the dawn of the nuclear age. If plans to build them faltered in the 1990s, they found renewed support in the 2000s, in part to shore up such strategic military enterprises as Sevmash in Severodvinsk, the Baltic Shipyards in St. Petersburg, and in a number of other powerful design institutes. Russia's nuclear shipbuilding industry found a striking way to convert the military atom into a peaceful atom. After the Cold War, such ship-building yards as Zvezdochka and Sevmash on the White Sea fell on hard times. Small businesses that spun off from the shipyards that produced decanters and glassware were hardly profitable, and were also an insult to the heroic defense traditions of the factories according to leaders and employees alike. Sevmash, with nearly 30,000 employees found better renaissance markets in manufacture of oil-drilling platforms, and there have been discussions of atomic powered platforms. But the major production will be nuclear attack submarines—as befitting a Putin-era ZATO.

Precisely floating nuclear power stations (PAES) were adored because they can provide electric energy and heat above the Arctic Circle, according to their promoters in almost any bay or inlet, and may be used to desalinate water for potential clients in a series of arid, equatorial countries. Rosatom intends to sell them for $335 million each: China, Algeria, Indonesia, Brazil, Malaysia, Indonesia, Mozambique, Namibia, South Africa, Egypt, Jordan, Kuwait, Vietnam, and others have expressed an interest in them. Orders seem to depend on whether the long-promised, very long delayed first PAES, the "Academic Lomonosov," operates efficiently at Pevek, on Chukotka's northern Arctic shore where it was moored in September 2019, twelve years after construction began in St. Petersburg. Mobile and floating reactors are nothing new, have completely military roots, and were developed in the United States as well.

The "Lomonosov," named after the eighteenth-century native son of Arkhangelsk province and revered Russian scientist, Mikhail Lomonosov, was launched from the Baltic Shipbuilding Yard in June 2017 for fueling in Murmansk, and later tugged to Pevek on the Eastern

Arctic Coast, where it was moored in a bay and surrounded by buffers against ice and waves. The hope is that the PAES will rejuvenate Pevek as a logistical center of the Northern Sea Route. In theory a PAES barge may hold four units, either power plants or desalination plants, or a combination of these applications, plus living quarters for sixty workers. The "Lomonosov's" two reactors, adapted from submarines, each produce 35 MWe with a hoped-for relatively repair-free lifetime of up to thirty-eight years. The PAES is equipped with single rooms, sports facility, and sauna, library, hair salon, and even a bakery reputed to produce the best bread in the region. Unfortunately, the float to Pevek took decades and the "Lomonosov" cost three times the original estimate.[13] Virtually all of the world's nuclear projects are well over cost and behind schedule.

The most majestic, heroic and, for Russians, nostalgic Arctic nuclear applications are icebreakers. Russia's nuclear icebreakers were dedicated to normalizing shipping, opening Arctic resources, and serving military bases. Like Stalin's public airplane spectacles to demonstrate enlightened leadership and Soviet achievement, so for Russia's current leaders, icebreakers are a kind of technological sublime.[14] They demonstrate control over nature and geography, national will, and world leadership at the cutting edge of engineering, and they have enabled assimilation of Arctic resources. Russian leaders, businessmen, journalists, and citizens have ignored such problems as massive cost overruns, lags in construction, and several accidents in running the icebreakers, instead celebrating them as a sign of national virility as the preeminent icebreaker power over Canada, the United States, and other nations. Russian specialists forecast that by 2030 they will have established year round transport to strategic sites along the Russian coast—through the construction of no fewer than forty new icebreakers, a number of them nuclear powered and run by Atomflot, a division of Rosatom. On many occasions, Putin has shown up in person or by video link to christen new ships or shipyards—or pipeline pumping stations or submarines—served by Arctic tankers.[15]

Physicists at KIAE and engineers of Sevmash have in mind precisely using submarine technology for atomic oil and gas exploration in treacherous arctic waters. KIAE specialists, together with those at the

All-Union Research Institute of the Oil and Gas Industry, determined that nuclear power will solve the "grandiose problem of the assimilation of the arctic shelf," by which they mean extracting oil and gas from what may be 30 percent of the world's reserves in conditions of ice and deep ocean. They propose autonomous underwater surveying, extracting, and transportation technology powered by nuclear energy that will be "energy efficient" and with the "smallest risk of negative impact on the environment." They have advanced such floating nuclear devices as tankers, compressor stations, underwater drilling equipment, and so on. Underwater tankers have been designed by the Malakhit Engineering. Another engineering group proposes an underwater LNG tanker at 277,000 T at 260 m in length, and perhaps $600 million in cost.[16]

Putin's Atom

Putin did not initially handle well his image as an all-powerful, masculine, military leader, determined to protect Russia from any danger. When the "Kursk" submarine sank on August 12, 2000, in the Barents Sea, killing all 118 personnel on board after a torpedo fuel explosion that was initially blamed on collision with a foreign boat, Putin continued his vacation at a Black Sea resort, and he authorized the Russian Navy to accept British and Norwegian offers of assistance only five days after the accident. This recalled the earlier Soviet treatment of accidents: to be silent. The Russians mishandled the accident and misled the public about what had occurred and what, if anything, they had done in the emergency response; a Norwegian operation to recover the bodies found that the sailors had lived sometime before suffocating or burning to death.[17] Fortunately the reactor was intact. Once the submarine was hauled into a Naval base, the fuel, roughly 1.2 tons of enriched uranium in the twin reactors, was removed and taken by train to Maiak.

I myself was fortunate to go to Severodvinsk on many occasions in the 2000s, as a member of the Portsmouth, New Hampshire-Severodvinsk sister city organization, before the authorities closed that

ZATO to foreigners again as a result of the government's increasingly hostile foreign policy toward the United States, the west, and NATO. I learned firsthand how important the military atom remains to the nation, even if accidents have taken lives of heroic sailors and nuclear waste has been a part of daily life. Russia built 248 nuclear submarines and five naval surface vessels (plus nine icebreakers) powered by 468 reactors between 1950 and 2003.[18]

Even today there are over forty ZATOs in Russia with one-million-plus residents that house Russia's military-industrial complex including its weapons production and waste facilities: Cheliabinsk-65 (now Ozersk, home of the major Soviet plutonium production facility), Tomsk-7 (now Seversk, home of a fuel reprocessing factory and of a major nuclear accident in 1993), Krasnoiarsk-26 (now Zheleznogorsk, home of another plutonium processing facility), Sarov (Arzamas-16, where bombs were designed), the ship-building center Molotovsk (today Severodvinsk, on the North Sea), and many others that indicate the historical importance of the atom in Russian culture and politics.

Putin will not make this mistake of being distant from the atom—or accidents—again. Putin instead has avidly pursued the nuclearization of the nation, in particular of the Arctic, with a return to Soviet-era nuclear doctrine that justifies first use of nuclear weapons if the situation determines it. This is reflected precisely in building and testing nuclear submarines and missiles, for example, the launch of the "Krasnoiarsk" submarine in July 2021 from Sevmash, like others to be built to carry cruise and ballistic missiles.[19] A promised hypersonic cruise missile, the Tsirkon (Zircon), for the first time from a submarine, followed soon thereafter. Putin announced "an array of new hypersonic weapons in 2018 in one of his most bellicose speeches in years, saying they could hit almost any point in the world and evade a U.S.-built missile shield."[20]

Russia remains prepared for nuclear war, even as the United States and Russia signed the New START treaty (April 2010) to limit offensive strategic weapons. According to data from New START, Russia deploys 1,444 strategic warheads on 527 ICBMs, SBLMs, and heavy bombers. Russia's arsenal includes five ICBMS, three of which were developed during the Soviet period, and two of which were developed in the Russian Federation. They include (by NATO designation): SS-18s, SS-19s, SS-25s (the Soviet ones), the SS-27 [the world's first mobile ICBM] and SS-29. They are MIRV-ed with between six and ten warheads, and Soviet missile service life has been extended by fifteen years several times. Russia is striving to retire its Soviet-era heavy ICBMs with many warheads and replace them with smaller solid fueled missiles with fewer warheads, and the total number of deployed warheads has decreased. New START required Russia to cut its strategic nuclear arsenal to 1,550 operational warheads and 800 deployed and non-deployed launchers by February 2018, and it did so.[21]

But since the early 2000s the treaties that limit arms or require cuts in numbers have been buffeted by a variety of challenges including continued misunderstandings among leaders and negotiators of the nuclear nations. From the US side, fears that Russia has continued to mislead the United States or taken advantage of existing treaties led the country unilaterally to take advantage of existing language with withdraw from several of them, in two major cases under Republican presidents at the urging of neo-conservatives who boast about strength and power, and who do not truly seek arms control. In 2002 George W. Bush abrogated the 1972 ABM treaty claiming that it prevented the United States from defending its own people from attack, and asserting that the Russians had violated aspects of it, but in fact to push ahead with the Strategic Defense Initiative ["Star Wars"] defense system that most practicing physicists believe has no possibility of operating as intended. Sixty billion dollars have been spent and there has not been a step toward the initial goal of rendering nuclear weapons obsolete and impotent as claimed by supporters.

And second, signed in 1987, the INF Treaty led to the elimination of 2,692 US and Soviet nuclear and conventional ground-launched ballistic and cruise missiles with ranges between 500 and 5,500 kilometers. But in September 2019 Donald Trump abrogated the INF treaty. As reported by the Arms Control Association, "For several years, the United States has alleged that Russia was in violation of the INF Treaty by testing and deploying a banned missile system, and Washington pinned its treaty withdrawal squarely on Russia. 'Russia is solely responsible for the treaty's demise', said Secretary of State Mike Pompeo in announcing the US move. 'Over the past six months, the United States provided Russia a final opportunity to correct its noncompliance. As it has for many years, Russia chose to keep its noncompliant missile rather than going back into compliance with its treaty obligations.'" Since 2014, the United States has accused Russia of violating the treaty by testing, possessing, and fielding an illegal ground-launched cruise missile (GLCM), known as the 9M729. The US DOD has begun to test its own GLCM.[22]

For its part, Russia, too, has become increasingly bellicose and assertive in its nuclear doctrine, in some respects returning to the rhetoric, if not the behavior, of Cold War years. This is not surprising given that Vladimir Putin is a product of the Cold War, a former KGB agent who sees foreign and domestic enemies working together to weaken Russia, and who believes that state power—and resource development—are the keys to Russian survival. He thus sees aggressive intent, not defense, in US nuclear weapons programs, and has determined to overhaul and modernize Russian forces, including nuclear forces. Increased budget outlays for new weapons programs are visible in all branches of the Russian military.

Putin displayed his rancor over the abrogation of the INF treaty. He said, "Now they are leaving the treaty on eliminating the short and middle-range missiles. What's next? It's hard to imagine how the situation will evolve. What if those missiles appear in Europe? What do we do then?" Indeed, a number of specialists insist that leaving the treaty will only trigger an arms race with ground-based nuclear missiles returning to Europe for the first time in decades. On top of this, according to new military doctrine, a recently published document reaffirms that the Russian could use nuclear weapons in response to a nuclear attack or an aggression involving conventional weapons that "threatens the very existence of the state." This kind of aggression could include the use of nuclear weapons or other weapons of mass destruction against Russia or its allies and an enemy attack with conventional weapons that threatens the country's existence.

In addition, Putin has asserted that Russia could use its nuclear arsenals if it gets "reliable information" about the launch of ballistic missiles targeting its territory or its allies and also in the case of "enemy impact on critically important government or military facilities of the Russian Federation, the incapacitation of which could result in the failure of retaliatory action of nuclear forces."[23] In November 2020 Putin said, "We don't want to burn bridges, but if somebody interprets our good intentions as weakness, our reaction will be asymmetrical, rapid and harsh … We'll decide for ourselves in each case where the red line is."[24] Unfortunately, the concept of preempting an enemy's nuclear weapons that was a central point of Soviet military thought has been reiterated by President Vladimir Putin. All the while, Russia wants to recreate several territorial aspects of the former Soviet empire, and through the annexation of Crimea, a war with Georgia and its unprovoked 2022 war in Ukraine Russia has only exacerbated mistrust and gained international condemnation.

Nuclear Russia in the Twenty-First Century: The Radiant Future?

From Mars to the moon with space-based reactors, from inlets and bays on the Arctic shore with PAES, to deep sea drilling in nuclear platforms, and from the usual PWRs and LMFBRs, Rosatom has established a powerful vision for the ubiquitous power of the atom (see Table 2, following). That power rests on nostalgia for the Soviet era, and the tremendous momentum of the institutes, factories, and design bureaus of the nuclear enterprise that have recovered from the shock of the end of the Cold War and the breakup of the USSR. It counts on the support of a resource state whose leaders see exploitation of oil and gas as the key to Russia's strength. And it moves forward with few checks on its programs. Some of the public is skeptical of these hubristic applications, but citizens have essentially been deprived of the right to protest against the nuclear enterprise—or to engage in any kind of protest—in Putin's Russia. Yet most of them accept post-Chernobyl nuclear power as safe and important to the nation's energy balance. Whether Rosatom has gained public acquiescence or simply can ignore what it believes are misplaced safety concerns, the nuclear enterprise can and will embrace the largesse of the government as it advances its programs and those of a nuclear-powered resource state.

Russia simultaneously determined to build up its military bases and available hardware in the High North that includes bombers and MiG31BM jets, and new radar systems near Alaska. A new weapon in its arsenal is the "Poseidon" nuclear-armed torpedo powered by a nuclear reactor and intended by Russian designers "to sneak past coastal defenses—like those of the US—on the sea floor." It would carry a warhead of many megatons, causing radioactive waves to make the target coastline uninhabitable for decades.[25] Goodman and Kertysova recently wrote that "the Russian Arctic will constitute the most nuclearised waters on the planet by 2035."[26] But Russia already created the most nuclearized waters in the world. It is nuclear in its nuclear dump sites, nuclear-powered icebreakers, and floating nuclear power stations. As we have now studied, it is perhaps the most aggressive nation of the world in pursuit of the peaceful and military atom. With physicists occupying nearly unassailable positions of prestige, and with political leaders sharing their religious firm belief that electricity—production of copious amounts of electricity—guarantees the empire's glorious future, the nuclear lobby gained broad budgetary and social support. Its generals never wanted for rubles, its physicists never worried if an ambitious, even far-fetched, program to power locomotives or airplanes or to use explosives to build dams would gain high-level support. Nuclear-powered submarine tankers and floating reactors—the offspring of these programs—have found support at the highest levels of government.

The effort to nuclearize the Arctic region has risks beyond the environmental costs of building infrastructure and the spread of waste as vividly made clear from a deadly nuclear accident involving cruise missile in 2019. Building on extensive Cold War infrastructure in and around the shipbuilding ZATO of Severodvinsk, the military began testing the Burevestnik nuclear-powered cruise missile. But test launches in the Barents Sea resulted in repeated failures that necessitated recovery efforts with special waste handling and underwater salvage vessels to bring radioactive material from the weapon's nuclear core back to the surface. A Chernobyl-like accident occurred on August 8,

2019, at the State Central Navy Testing Range in Nenoksa on the White Sea shoreline but 45 kilometers from Severodvinsk (pop. 200,000) and 70 km from Arkhangelsk (pop. 350,000) when the "isotope power source" for a liquid-fueled rocket engine exploded, although there is disagreement over the exact event and its cause. Local fisherman saw and heard the explosion and gave conflicting accounts of what happened. Reportedly, seven individuals were exposed to radiation and died in the explosion, or died later and were transported to local hospitals where the hospital staff were not warned of the radiation risk and none wore radiation protective equipment. Some of the medical personnel and victims were flown to Moscow for radiation testing where medical staff were forced to sign non-disclosure agreements. The Russian meteorological service stated that background radiation levels peaked at four to sixteen times normal levels at six of its eight stations in Severodvinsk. Fearing the worst, local residents scurried to pharmacies in hopes of finding iodine tablets.[27] Just like for Chernobyl, in this accident the authorities failed to inform the public about the risks they faced and how to protect themselves and their families.

In November 2019, Putin held a press conference to provide some details of the accident in remarks that shed little light on the accident itself, but in which he maintained, like a Cold War Warrior, that nuclear weapons ensured peace not war. Putin said, "First of all, it is my duty, but still an honorable mission, to express to you deep sincere condolences" from the Russian state. He then presented awards to the families of those killed in the tests near Severodvinsk. Those who died had chosen a "special mission to defend Russia, and they honorably fulfilled their duty to the end," and they "sought to ensure peace for future generations through the development of advanced weapons." Putin continued that each of the dead had "made an indispensable contribution to the strengthening of the Russian state," and he stressed that the mere fact of possessing such unique technologies was "a reliable guarantee of peace in the world." No matter what, Putin concluded, Russia would improve its nuclear weapons.[28]

Rosatom itself is critically important to Russia's geopolitical interests, economic power, and nostalgic self-image as a scientific superpower. According to national leaders, Rosatom, the inheritor of civilian and military programs of the Soviet nuclear ministry, "MinSredMash," will protect Arctic regions from growing foreign competition. It will assist in developing extensive national resources, and perhaps complete the unfinished tasks of the legendary northern shipping route that Stalin's super commissariat, Glavsevmorput, established. Russia's look to the Arctic as a site of both great economic value in its oil, gas, and other resources and of potential military conflicts requires the atom in two ways: first to power resource development through a variety of power generation and transport applications and through expansion of Artic military capabilities, already thoroughly nuclear in the submarines of the Northern Fleet and in the Novaia Zemlia testing ground.

Virtually all observers hope that the nuclear nations of the world will continue to pursue meaningful steps toward disarmament and greater cooperation in developing the peaceful atom. Yet from 1945 the belligerent nations of the world built 125,000 nuclear warheads, nearly all of them (about 97 percent) by the United States and the Soviet Union and Russia. Even in 2021 there are 13,000 in the world, with the lion's share—4,000 each—in Russia and the United States, more than enough to destroy the world directly or indirectly through "nuclear winter," the cooling of the earth resulting from dust, particulate and debris that rises into the atmosphere from nuclear holocaust that prevents the sunlight from getting through.[29]

The peaceful atom—to temper the military atom and to produce electricity—has encountered challenges and threats, too. In terms of challenges, the industry faces growing costs and delays in meeting production targets, uncertainty what to do with high level radioactive waste including SNF, and the potential for accidents. As alluded to, in France, England, Finland, and Russia cost overruns and safety and construction delays have belabored efforts to bring reactors on line. The Fukushima disaster of March 2011 that resulted in three reactor meltdowns and extensive radiological pollution indicates the ever-present risk of catastrophic disaster remains. Russian officials and engineers brushed off Fukushima as something that could not happen in Russia. They continue to believe the peaceful atom is safe, and that the military atom is required for defense. This is not surprising as this exploration of the historical roots of the unwavering political, economic, and ideological support of the atom and the neutron in Russia over the past 100 years has demonstrated.

Nuclear Russia tried to put Chernobyl—and its radiation that will continue to plague some 50,000 km^2 of Russia, Belarus and Ukraine for thousands of years—into the long past. After the Chernobyl disaster in 1986, leading specialists worried openly about continued operation of the remaining Chernobyl-type RBMK reactors at the Kursk, Smolensk, and Sosnovy Bor (Leningrad) sites, and eventually decided to kept those stations open, if on a different operating regime.

But Russia's rich and glorious, yet costly, controversial and dangerous embrace of the atom took a dreadful turn during Putin's unprovoked attack on Ukraine in February 2022, including attacks on Chernobyl. During the murderous war, Russia soldiers specifically targeted Ukrainian nuclear power stations, research institutes, and waste storage facilities. Russia, which had claimed to be the first nation of the peaceful atom, had become a nuclear aggressor. Russian soldiers entered the Chernobyl exclusion zone on February 24, their vehicles raising dust and causing spikes in radioactivity. Greatly increasing radiation dangers, the soldiers built foxholes in the "Red Forest," a 10 km^2 region of pine trees that died after the 1986 accident, and was bulldozed and buried. Russian military planners have targeted other Ukrainian nuclear facilities including waste dumps and power stations, including a radioactive waste burial facility in Kyiv with missiles. They shelled the UFTI. The Russian destruction of the institute, including already a neutron generator, is an obvious effort to handicap future Ukrainian research—and could lead to the release radiation in a city of 1.4 million inhabitants. After several days of battle in which unarmed residents and Ukraine's national guard fought off superior forces, Russian columns moved into the Zaporizhzhya NPP in Enerhodar. Interfering with safe operation and damaging buildings and power lines on site, Russia risks a nuclear disaster greater than Chernobyl. The Ukrainian government rightly called this unprovoked and unjustified military aggression an act of state-sponsored nuclear terrorism.[30] Under Vladimir Putin Russia has fully abandoned Igor Kurchatov's 1950s admonition for the "atom to be a worker, not a soldier."

Table 2 2021 Present Russian Nuclear Power Capacity Power reactors in operation

Reactor	Type V=PWR	MWe net, each	Commercial operation	Licensed to, or scheduled close
Akademik Lomonosov 1	KLT-40S	32	05/20	2029
Akademik Lomonosov 2	KLT-40S	32	05/20	2029
Balakovo 1	V-320	950	5/86	2043
Balakovo 2	V-320	950	1/88	2033
Balakovo 3	V-320	950	4/89	2049
Balakovo 4	V-320	950	12/93	2053
Beloyarsk 3	BN-600 (FBR)	560	11/81	2030
Beloyarsk 4	BN-800 (FBR)	820	10/16	2056
Bilibino 2-4	EGP-6 (LWGR)	3 x 11	12/74-1/77	Dec 2021; unit 2: 2025
Kalinin 1	V-338	950	6/85	2045
Kalinin 2	V-338	950	3/87	2047
Kalinin 3	V-320	950	11/2005	2065
Kalinin 4	V-320	950	9/2012	2072
Kola 1	V-230	411	12/73	2033
Kola 2	V-230	411	2/75	2034
Kola 3	V-213	411	12/84	2027
Kola 4	V-213	411	12/84	2029
Kursk 1	RBMK	925	10/77	2022
Kursk 2	RBMK	925	8/79	2024
Kursk 3	RBMK	925	3/84	2029
Kursk 4	RBMK	925	2/86	2031
Leningrad 3	RBMK	925	6/80	2025
Leningrad 4	RBMK	925	8/81	2026
Leningrad II-1	V-491	1101	10/2018	2078
Leningrad II-2	V-491	1066	03/2021	2079

Reactor	Type V=PWR	MWe net, each	Commercial operation	Licensed to, or scheduled close
Novovoronezh 4	V-179	385	3/73	2032
Novovoronezh 5	V-187	950	2/81	2035 potential
Novovoronezh II-1*	V-392M	1100	10/2018	2077
Novovoronezh II-2*	V-392M	1101	03/2021	2077
Rostov 1	V-320	950	3/2001	2031
Rostov 2	V-320	950	10/2010	2040
Rostov 3	V-320	950	9/2015	2045
Rostov 4	V-320	979	9/2018	2048
Smolensk 1	RBMK	925	9/83	2028
Smolensk 2	RBMK	925	7/85	2030
Smolensk 3	RBMK	925	1/90	2034
Total: 38		**28,578 MWe**		

The V-320 is the base model of what is generically VVER-1000; V-230 and V-213 are generically VVER-440; V-179 & V-187 are prototypes. Rostov is sometimes known as Volgodonsk. Most closure dates are from January 2015 unless license extension indicates later date. Many reactors have been uprated but current net capacities are mostly unknown. At the end of 2016 all 11 VVER-1000 reactors were operating at 104 percent of original capacity.

Source: Nuclear Power in Russia (updated August 2021), World Nuclear Organization at https://world-nuclear.org/information-library/country-profiles/countries-o-s/russia-nuclear-power.aspx

NOTES

Introduction

1 After the breakup of the USSR, Ukraine held about one-third of the Soviet nuclear arsenal, the third largest number in the world at the time, and extensive design and production capability. But in 1994 Ukraine agreed to destroy the weapons, and sign the NPT. See Robert Norris "The Soviet Nuclear Archipelago," *Arms Control Today*, vol. 22, no. 1 (January–February 1992): pp. 24–31; and William C. Martel, "Why Ukraine Gave Up Nuclear Weapons," in Barry R. Schneider, and William L. Dowdy, eds., *Pulling Back from the Brink: Reducing and Countering Nuclear Threats* (London: Frank Cass, 1998), pp. 88–104.

Chapter 1

1 On the history of Ukrainian physics, from which this section is drawn in part, see Paul Josephson, Yuri Ranyuk, Ivan Tsekhmistrov and Karl Hall, "Science and the Periphery Under Stalin: Physics in Ukraine," in Helmuth Trischler and Mark Walker (eds.), *Physics and Politics* (Stuttgart: Franz Steiner Verlag, 2010), pp. 197–226. See also Iu.V. Pavlenko, Iu.A. Ranyuk and Iu.A. Khramov, *"Delo" UFTI, 1935–1937* (Kiev: Feniks, 1998).

2 On the famine in Ukraine, Robert Conquest, *Harvest of Sorrow* (New York: Oxford, 1986).

3 GAKhO (State Archive of Kharkiv Oblast), F. 17-2, op. 1, d. 325, str. 52, and F. 17–23, op. 1, d. 3, str. 141.

4 GAKhO, F. p-23, op. 1, d. 7, str. 10–12. Landau fled to the Peter Kapitsa's Institute of Physical Problems in Moscow. The secret police caught up with him in April 1938, arrested, interrogated, and tortured him, freeing him at near death only after the personal intervention of Peter Kapitsa after one year in prison.

5 GAKhO, F. p-23, op. 1, d. 7, str. 10–12.

6 GAKhO, F. p-23, op. 1, d. 10, str. 146–147.

7   Robert McCutcheon, "The 1936–37 Purge of Soviet Astronomers,"
     Slavic Review, vol. 50, no. 1 (1991): pp. 100–17; and A. I. Eremeeva,
     "Political Repression and Personality: The History of Political
     Repression against Soviet Astronomers," *Journal for the History of
     Astronomy*, vol. 26, no. 4 (1995): pp. 297–324.

8   Robert Forczyk, *Kharkov 1942: The Wehrmacht Strikes Back* (Oxford:
     Osprey Press, 2009) and David Glantz, *Kharkiv 1942: Anatomy of a
     Military Disaster* (New York City: Sarpedon, 1998).

9   "Drobitskii Iar Na Vostochnoi Okraine Gorodo Khar'kova," at http://
     www.drobytskyyar.org/. See also A. V. Skorobogatov, *Kharkiv u Chasi
     Nimets'koi Okupatsi (1941–1943)* (Kharkiv: Prapor, 2004).

10   A KIAE (Archive of Kurchatov Institute of Atomic Energy), F. 2, op. 1,
     d. 71/8, ll. 1–4. See also Stephen Wheatcroft, "The Scale and Nature of
     German and Soviet Repression and Mass Killings, 1930–45," *Europe-
     Asia Studies*, vol. 48, no. 8 (1996): pp. 1319–53.

11   A KIAE, F. 1, op. 1/c, ed. khr. 63.

12   Letter from Sinelnikov to Beria, February 29, 1952, A KIAE, F. 1, op.
     1/c, d. 78.

13   GAKhO, F. 854, op. 1, d. 68, ll. 116–120.

Chapter 2

1   "Statement by President Truman in Response to First Soviet Nuclear
     Test," September 23, 1949, History and Public Policy Program Digital
     Archive, Department of State Bulletin, Vol. XXI, No. 533, October 3,
     1949, at https://digitalarchive.wilsoncenter.org/document/134436.pdf.

2   The best study of the Soviet atomic bomb project is David Holloway, *Stalin
     and the Bomb: The Soviet Union and Atomic Energy, 1939–1956* (New
     Haven: Yale University Press, 1994). See also the three-volume collection of
     documents and others materials on the history of the Soviet atomic bomb
     project, completed at over 8,000 pages and with a detailed guide, *Atomnyi
     Proekt SSSR. Dokumenty i Materialy* (see online http://elib.biblioatom.ru/
     sections/0201/). See also Robert Norris, Thomas B. Cochran, "Nuclear
     Weapons Tests and Peaceful Nuclear Explosions by the Soviet Union:
     August 29, 1949, to October 24, 1990," Natural Resource Defense Council,
     May 19, 2013, at https://nuke.fas.org/norris/nuc_10009601a_173.pdf. On
     espionage, see Jeremy Bernstein, "John von Neumann and Klaus Fuchs:
     an Unlikely Collaboration," *Physics in Perspective*, vol. 12, no. 1 (March
     2010): pp. 36–50; Dieter Hoffmann, "Fritz Lange, Klaus Fuchs, and the
     Remigration of Scientists to East Germany," *Physics in Perspective*, vol. 11,
     no. 4 (December 2009): pp. 405–25.

Notes

3   On ZATOs, Asif Siddiqi, "Atomized Urbanism: Secrecy and Security from the Gulag to the Soviet Closed Cities," *Urban History*, (2021): pp. 1–21; Michael Gentile, "Former Closed Cities and Urbanisation in the FSU: An Exploration in Kazakhstan," *Europe-Asia Studies*, vol. 56, no. 2 (March 2004): pp. 263–78; and Richard H. Rowland, "Russia's Secret Cities," *Post-Soviet Geography and Economics*, vol. 37, no. 7 (1996): pp. 426–62.

4   Atomic Heritage Foundation, "Georgy Flerov," https://www.atomicheritage.org/profile/georgy-flerov.

5   Peter Kelly, "How the USSR Broke into the Nuclear Club," *New Scientist*, vol. 110 (May 8, 1986): pp. 32–6.

6   Clarence Lasby, *Project Paperclip: German Scientists and the Cold War* (New York: Atheneum, February 1971); and Samuel Goudsmit, *Alsos* (New York: Henry Schuman, 1947).

7   *V Mire Nauki*, no. 4 (2013) at file:///Users/prjoseph/Downloads/1-2013-v_mire_nauki-special-issue-4-2013.pdf

8   For declassified records of the history of KIAE, I. V. Kurchatov, *Sobranie Nauchnykh Trudov. V Shesti Tomov* (Moscow: Nauka, 2005–2012).

9   "LIPAN" at http://46.29.163.254/wiki/%D0%9B%D0%98%D0%9F%D0%90%D0%9D

10   Amy Knight, *Beria: Stalin's First Lieutenant* (Princeton, NJ: Princeton University Press, 1993).

11   G. Ozerov, *Tupolevskaia Sharaga* (Moscow: Posev, 1971).

12   D. Granin, *Zubr* (Leningrad: Sovetskii Pisatel', 1987) at http://lib.ru/PROZA/GRANIN/zubr.txt. For a defense of Timofeev-Resovsky of charges that he supported the Nazi eugenics programs, Raissa Berg, "In Defense of Timoféeff-Ressovsky," *The Quarterly Review of Biology*, vol. 65, no. 4 (1990): pp. 457–79.

13   Nikolaus Riehl and Frederick Seitz, *Stalin's Captive: Nikolaus Riehl and the Soviet Race for the Bomb* (New York: Chemical Heritage Foundation, 1996); and Pavel Oleynikov, "German Scientists in the Soviet Atomic Project," *The Nonproliferation Review*, vol. 7, no. 2 (2000): pp. 1–30.

14   Jonathan Brent and Vladimir P. Naumov, *Stalin's Last Crime: The Plot against the Jewish Doctors, 1948–1953* (New York: Harper Collins, 2003); Roy Medvedev, *Let History Judge: The Origins and Consequences of Stalinism* ed. and trans. George Shriver, rev. and exp. ed. (New York: Columbia University Press, 1989); Iakov L'vovich Rapoport, *The Doctors' Plot of 1953*, trans. N. A. Perova and R. S. Bobrova (Cambridge: Harvard University Press, 1991).

15   Henry Smyth, *Atomic Energy for Military Purposes. The Official Report on the Development of the Atomic Bomb under the Auspices of the*

United States Government 1940–1945 (Princeton: Princeton University Press, 1945).

16 On Khariton, see Holloway, p. 94; and German Goncharov and Lev Ryabev, "The Development of the First Soviet Atomic Bomb," *Physics-Uspekhi*, vol. 44, no. 1 (2001): pp. 71–93. See also Oleg Bukharin et al., *Russian Strategic Nuclear Forces* (Cambridge: MIT Press, 2001).

17 Wikipedia, "Soviet Atomic Bomb Project," https://en.wikipedia.org/wiki/Soviet_atomic_bomb_project; and Steve Zaloga, *The Kremlin's Nuclear Sword: The Rise and Fall of Russia's Strategic Nuclear Forces* (Washington, DC: Smithsonian Books, 2002), pp. 32–5.

18 Comprehensive Nuclear Test Ban Treaty Organization (CTBTO), "The Soviet Union's Nuclear Testing Programme," at https://www.ctbto.org/nuclear-testing/the-effects-of-nuclear-testing/the-soviet-unionsnuclear-testing-programme/.

19 T. Artemov and A. E. Bedel, *Ukroshchenie Urana* (Ekaterinburg: Izdatel'stvo 000 SV 96,1999), pp. 252–5.

20 Artemov and Bedel, *Ukroshchenie Urana*, pp. 131, 138, 195, 211–12. At the end of the 1950s and early 1960s Soviet capacities for gaseous diffusion expanded greatly at the new Siberian Chemical Combine at Tomsk-7, and the Angarsk Electrolytic Chemical Combine and the Electrochemical Factory at Krasnoiarsk-45. This required production of fantastical numbers of filters which, in 1961, reached 66.7 million, and in thirty years 1 billion of them with total length of 500,000 km.

21 NTI, "Semipalatinsk Test Site," *Nuclear Threat Initiative*, https://www.nti.org/learn/facilities/732/; CTBTO, "The Soviet Union's Nuclear Testing Programme"; Norris and Cochran, "Nuclear Weapons Tests and Peaceful Nuclear Explosions by the Soviet Union"; and Peter Stegner and Tony Wrixon, "Semipalatinsk Revisited," *IAEA Bulletin*, vol.40-4 (December 1998): pp. 12–14, at https://www.iaea.org/sites/default/files/publications/magazines/bulletin/bull40-4/40405081214.pdf.

22 Ibid., NTI, "Semipalatinsk Test Site"; and Norris and Cochran, "Nuclear Weapons Tests."

23 B. I. Gusev, Z. N. Abylkassimova, K. N. Apsalikov, "The Semipalatinsk Nuclear Test Site: A First Assessment of the Radiological Situation and the Test-related Radiation Doses in the Surrounding Territories," *Radiation and Environmental Biophysics*, vol. 36, no. 3 (September 1997): pp. 201–4.

24 L. E. Peterson et al., "Diagnosis of Benign and Malignant Thyroid Disease in the East Kazakhstan Region of the Republic of Kazakhstan," *Cancer Research Therapy and Control*, vol. 4, no. 4 (1998): pp. 307–12; R. Vakulchuk et al., "Semipalatinsk Nuclear Testing: The Humanitarian

Consequences," Norwegian Institute of International Affairs, *Report No. 1,* February 2014; Gusev, Abylkassimova, Apsalikov, "The Semipalatinsk Nuclear Test Site": S. L. Simon, K. F. Baverstock and C. Lindholm, "A Summary of Evidence on Radiation Exposures Received Near to the Semipalatinsk Nuclear Weapons Test Site in Kazakhstan," *Health Physics,* vol. 84, no. 6 (2003): pp. 718–25.

25   US House of Representatives, Subcommittee on Energy Conservation and Power, *American Nuclear Guinea Pigs: Three Decades of Radiation Experiments on US Citizens* (Washington, DC: USGPO, 1986)

26   CTBTO, "The Soviet Union's Nuclear Testing Programme"; Judith Perrara, "Forty Year Old Secret Explodes …," *New Scientist,* Issue 1904 (December 18, 1993), at https://www.newscientist.com/article/mg14019041-400-forty-year-old-secret-explodes/.

27   Original Soviet film: "Atomnyi Vzryv—Totskie Ucheniia," at https://www.youtube.com/watch?v=GZSc7I13sPU. See also Atomic Heritage Foundation, "Operation Plumbbob—1957," at https://www.atomicheritage.org/history/operation-plumbbob-1957.

28   "Reshaiushchii Argument: Kak Sovetskaia 'Tsar'-Bomba' Pomogla Sokhranit' Mir," *RIA,* October 30, 2016, at https://ria.ru/20161030/1480296569.html.

29   "Nashi Poliarniki. Matiushenko Tat'iana Pavlovna," *Sevmeteo,* June 2, 2020, at http://www.sevmeteo.ru/press/polars/8597/?sphrase_id=26708

30   Nikolai Biriukov, "Arkhipelag Osobogo Naznacheniia," *Vozdushno-Kosmicheskaia Oborona,* October 26, 2010 at http://www.vko.ru/author/nikolay-biryukov; Vitaly Khalturin et al., "A Review of Nuclear Testing by the Soviet Union at Novaya Zemlya, 1955—1990," *Science and Global Security,* vol. 13, nos. 1–2 (2005): pp. 1–42; Oleg Bukharin, *Russian Strategic Nuclear Forces*; *Iadernye Ispytaniia SSSR. Novozemelskii Poligon: Obespecheniye Obshchei i Radiatsionnoi Bezopasnosti Iadernykh Ispyta*nii (Moscow: IzdAT, 2000), pp. 75–8; and Nuclear Threat Initiative, "Central Test Site of Russia on Novaya Zemlya," at https://www.nti.org/learn/facilities/924/.

31   "Iadernoe Ispytanie no. 43 (1957)," at https://ru.googl-info.com/4248247/1/yadernoe-ispytanie-43-1957.html

32   Khalturin et al., "A Review of Nuclear Testing"; G. I. Ivanov, "Guba Chernaia: Chto Ostalos' Posle Vzryvov," *Besopasnost' Okuzhaiushchei Sredy,* 2007, at https://www.atomic-energy.ru/articles/2008/12/18/1196; V. V. Adushkin, B. D. Khristoforov, "Iadernye Vsryvy na Akvatorii Guby Chernoi," in *Iadernye Ispytaniia v Arktike,* Book 1, part 2 (Moscow, 2006), pp. 297–304, at http://elib.biblioatom.ru/text/yadernye-ispytaniya_kn1_t2_2006/go,296/

33   Tat'iana Kachalova, "'Tsar'-Bomba' – Mig Mezhdu Proshlym i

Budushchim," *Administratsiia Sankt-Peterburg*, June 14, 2018, at
https://www.gov.spb.ru/gov/terr/reg_viborg/news/138308/.

34 "Tsar'-Bomba: Ispytanie Samogo Moshchnogo Oruzhiia za Vsiu
    Istoriiu Chelovechestva," March 1, 2017, at https://knowhow.pp.ua/
    tsar_bomb/.

35 Kachalova, "'Tsar'-Bomba.'"

36 Kachalova, "'Tsar'-Bomba.'"

37 Timur Alimov, "Zemlia Drognula Trizhdy," *Russkaia Gazeta*, October
    30, 2019, at https://rg.ru/2019/10/30/zemlia-drognula-trizhdy-kak-car-
    bomba-spasla-mir-ot-novoj-vojny.html.

38 "Rassekrecheny Dannye ob Ispytanniiakh 'Tsar'-Bomby," *Lenta*,
    August 28, 2020, at https://lenta.ru/news/2020/08/28/tsar_bomba/;
    and Alimov, "Zemlia Drognula Trizhdy" for the specious claim that
    this monster saved the world from a new war.

39 George Quester, "Soviet Policy on the Nuclear Non-Proliferation
    Treaty," *Cornell International Law Journal*, vol. 5, no. 1 (1972), Article
    2, at https://scholarship.law.cornell.edu/cilj/vol5/iss1/2. Accessed April
    28, 2022.

40 FAS, "Russian/Soviet Doctrine," at https://nuke.fas.org/guide/russia/
    doctrine/intro.htm.

41 FAS, "Russian/Soviet Doctrine."

Chapter 3

1 For a discussion of the postwar cult of science, and several of these
   technologies, see Josephson, Paul, "Rockets, Reactors and Soviet Culture,"
   in Loren Graham, ed., *Science and the Soviet Social Order* (Cambridge:
   Harvard University Press, 1990), pp. 168–91.

2 In a section of "Children, Nation and Reactors: Imagining and
   Promoting Nuclear Power in Contemporary Ukraine," *Centaurus*, vol.
   61, no. 1 (2019): pp. 51–69, Tatiana Kasperski analyzes the evolution of
   covers of Soviet popular science magazines, and their devotion to space
   topics and less to the atom.

3 United Nations, Archival Materials, Folder S-1057-0008-03, Atomic
   Energy, First international Conference on Peaceful Uses of Atomic
   Energy, Geneva, 8-20 August 1955, at https://search.archives.un.org/
   atomic-energy-first-international-conference-on-peaceful-uses-of-
   atomic-energy-geneva-8-20-august-1955.

4 First International Conference on Atomic Energy Geneva, August 8–20,
   1955, *Bulletin of the Atomic Scientists*, vol. 11, no. 8 (October 1955):
   pp. 274–88.

5   Kurchatov, *Iadernaia Energetika*, vol. III (Moscow: Nauka, 1984),
     pp. 166–71.

6   I. N. Golovin, "Mirnyi Atom: U Istokov Sotrudnichestva," *Priroda*, no.
     5 (1988): pp. 122–8.

7   For a list of the various prototypes built and proposed, see Rosatom,
     "Istoriia Reaktorov," at http://www.biblioatom.ru/evolution/istoriya-
     osnovnyh-sistem/istoriya-reactorov/

8   On the ARBUS and its "minuses," see "Dvulikii 'ARBUS," at http://www.
     biblioatom.ru/evolution/istoriya-osnovnyh-sistem/istoriya-reactorov/arbus/

9   Oleg Bukharin, "Russia's Nuclear Icebreaker Fleet," *Science and Global
     Security*, vol. 14 (2006): pp. 25–31.

10   M. G. Meshcheriakov, *K 90-Letiiu so Dnia Rozhdeniia* (Dubna:
     OIIaI, 2000), pp. 52–7. On the importance of Soviet-Eastern Europe
     cooperation, see Sonja Schmid, "Nuclear Colonization?: Soviet
     Technopolitics in the Second World," in Gabrielle Hecht, ed., *Entangled
     Geographies: Empire and Technopolitics in the Global Cold War*
     (Cambridge: MIT Press, 2011), pp. 125–54.

11   George Ginsburgs, "Soviet Atomic Energy Agreements," *International
     Organization*, vol. 15, no. 1 (1961): pp. 49–65.

12   Alice Kimball Smith, *A Peril and a Hope: The Scientists' Movement in
     America 1945–47* (Chicago: University of Chicago Press, 1965).

13   For example, Alison Kraft, Carola Sachse, eds., *Science, (Anti-)
     Communism and Diplomacy: The Pugwash Conferences on Science
     and World Affairs in the Early Cold War*, Series: History of Modern
     Science, Volume 3 (Leiden and Boston: Brill, 2020). See also, Matthew
     Evangelista, *Unarmed Forces: The Transnational Movement to End the
     Cold War* (Ithaca: Cornell University Press, 1999).

14   B. I. Bulatov, *200 Iadernykh poligonov SSSR* (Novosibirsk: Tseris, 1993).

15   A. D. Sakharov, "Radioactive Carbon from Nuclear Explosions
     and Nonthreshold Biological Effects" from *Atomnaia Energiia*,
     vol. 4 (June 1958): pp. 576–580 in English translation at /https://
     scienceandglobalsecurity.org/archive/sgs01sakharov.pdf.

16   American Institute of Physics, "Nuclear Testing and Conscience" at
     https://history.aip.org/exhibits/sakharov/nuclear-testing.html.

17   United Nations, "Treaty on the Non-Proliferation of Nuclear
     Weapons," at https://www.un.org/disarmament/wmd/nuclear/npt/text/.

18   CTBTO, "General Overview of the Effects of Nuclear Testing," at
     https://www.ctbto.org/nuclear-testing/the-effects-of-nuclear-testing/
     general-overview-of-theeffects-of-nuclear-testing/. See also Rebecca
     Strode, "Soviet Policy toward a Nuclear Test Ban, 1958–1963," in
     Michael Mandelbaum, ed., *The Other Side of the Table: The Soviet
     Approach to Arms Control* (New York and London: Council on Foreign
     Relations Press, 1990), pp. 5–40.

Chapter 4

1   Robert Darst, *Smokestack Diplomacy: Cooperation and Conflict in East-West Environmental Politics* (Cambridge: MIT Press, 2001).

2   On impacts on local indigenes of nuclear testing around the world, see among many other studies: Lorna Arnold and M. Smith, *Britain, Australia and the Bomb: The Nuclear Tests and their Aftermath* (Basingstoke; New York: Palgrave, 2006); Bengt Danielsson and Marie-Therese Danielsson, *Poisoned Reign: French Nuclear Colonialism in the Pacific* (Ringwood: Penguin Books,1986); and Dorothy Nelkin, "Native Americans and Nuclear Power," *Science, Technology and Human Values*, vol. 6 (1981): pp. 2–12.

3   Alla Cherednichenko "Kto Razogreet Merzlotu," *Sankt-Peterburgskie Vedomosti*, August 3, 2017, at https://spbvedomosti.ru/news/obshchestvo/kto_razogreet_merzlotu/. Accessed April 28, 2022.

4   See Alexey Dudarev, Valery Chupakhin and Jon Øyvind Odland, "Health and Society in Chukotka: An Overview," *International Journal of Circumpolar Health*, vol. 72 (2013): pp. 1–10.

5   Vitalii Dragunov, "Litsa Goda," *Vestnik Atomprom*, September 30, 2013, at https://atomvestnik.ru/2013/09/30/lica-goda/. Accessed April 28, 2022.

6   Peter Bossew et al., "Radiological Investigations in the Surroundings of Bilibino, Chukotka, Russia," *Journal of Environmental Radioactivity*, vol. 51, no. 3 (December 2000): pp. 299–319.

7   Charles Digges, "Russia's Bilibino Nuclear Station Shutting Down Reactors to Make Way for Floating Nuclear Plant," *Bellona*, May 9, 2016, at https://bellona.org/news/nuclear-issues/2016-05-russias-bilibin-nuclear-station-shutting-down-reactors-to-make-way-for-floating-nuclear-plant. Accessed April 28, 2022.

8   Charles Digges, "Russia Set to Decommission the World's Most Remote Nuclear Power Plant," February 12, 2019, at Bellona at https://bellona.org/news/nuclear-issues/2019-02-russia-sets-to-decommission-the-worlds-most-remote-nuclear-power-plant. Accessed April 28, 2022.

9   Scott Kirsch, *Proving Grounds: Project Plowshare and the Unrealized Dream of Nuclear Earthmoving* (New Brunswick, NJ and London: Rutgers University Press, 2005). See on a far-fetched and dangerous proposed PNE to build an Alaska harbor, Dan O'Neill, *The Firecracker Boys* (New York: Basic Books, 2005).

10   On the extent of the Soviet Cold War legacy, see Aleksei Iablokov, et al., *Fakty i Problemy Sviazannye so Sbrosom Radioaktivnykh Otkhodov v Moria, Primykaiushchie k Territorii Rossiiskoi Federatsii* (Moscow: Priemnaia Prezidenta Rossiiskoi Federatsii, 1993). A series

of Bellona Foundation reports are vital: Nils Bøhmer et al., *The Arctic Nuclear Challenge* (Oslo, Norway: Bellona Foundation, 2001); and Igor Kudrik et al., *The Russian Nuclear Industry—The Need for Reform* (Oslo, Norway: Bellona Foundation, 2004).

11 Olga Bugrova, "Ne Pei Voditsu, Mutantonm Stanesh': Kak v SSSR Delali Ozera," *Radiosputnik*, January 15, 2021, at https://radiosputnik. ria.ru/20210115/chagan-1593021557.html. For the test explosion at Chagan, see "Chagan. Atomic Lake," December 20, 2009, at https:// www.youtube.com/watch?v=ZAoSUIASET0.

12 "Four Decades of Nuclear Testing: The Legacy of Semipalatinsk," *EClinicalMedicine*, vol. 13 (August 29, 2019), p. 1, at https://www. thelancet.com/action/showPdf?pii=S2589-5370%2819%2930151-8. See also Antoine Blua, "Effects Of 'Peaceful' Nuclear Tests Felt Decades Later," September 21, 2009, at https://www.rferl.org/a/Effects_Of_ Peaceful_Nuclear_Tests_Felt_Decades_Later/1827848.html.

13 Milo Nordyke, *The Soviet Program for Peaceful Uses of Nuclear Explosions* (Livermore: Livermore National Laboratory, 1996); and A. V. Iablokov, *Mif o Bezopasnosti i Effektivnosti Mirnykh Podzemnykh Iadernykh Vsryvov* (Moscow: TsEPR, 2003).

14 "Proekt 'Dnepr-1'. Kak Eto Bylo," at http://discoverkola.com/proekt-dnepr-1-kak-eto-bylo

15 Alexey Pavlov, trans. Maria Kaminskaya, "Former Nuclear Blast Sites in Russia's Murmansk Region to Become a National Park," *Bellona*, January 23, 2011, at https://bellona.org/news/future-energy-system/2011-01-former-nuclear-blast-sites-in-russias-murmansk-region-to-become-a-national-park.

16 See findings of Stephen Schwartz, ed., *Atomic Audit: The Costs and Consequences of US Nuclear Weapons since 1940*, at https://www. brookings.edu/the-hidden-costs-of-our-nuclear-arsenal-overview-of-project-findings/.

17 ICAN, "Complicit: 2020 Global Nuclear Weapons Spending," https:// www.icanw.org/complicit_nuclear_weapons_spending_increased_ by_1_4_billion_in_2020#:~:text=%2472.6%20billion%20is%20 how%20much,nuclear%20weapons%20took%20full%20effect.

18 Robert S. Norris, Hans M. Kristensen, "Nuclear U.S. and Soviet/ Russian Intercontinental Ballistic Missiles, 1959–2008," *Bulletin of the Atomic Scientists*, vol. 65, no. 1 (2009): p. 66; and Kristensen and Norris, "Global Nuclear Weapons Inventories, 1945–2013," *Bulletin of the Atomic Scientists*, vol. 69, no. 5 (2013): pp. 75–81.

19 Alan McDonald, "Scientific Cooperation as a Bridge across the Cold War Divide: The Case of the International Institute for Applied Systems Analysis (IIASA)," *Annals of the New York Academy of*

Sciences, vol. 866, no. 1 (December 1998): pp. 55–83. See also Eglė Rindzevičiūtė, *The Power of Systems: How Policy Sciences Opened Up the Cold War World* (Ithaca: Cornell University Press, 2016).

20   Nixon and Brezhnev met again in June 1973 in Washington, June 1974 in Moscow, and President Gerald Ford with Brezhnev in November 1974 in Vladivostok and in July 1975 in Helsinki to sign the Helsinki Accords.

21   Stephen Millett, "Forward-Based Nuclear Weapons and SALT I," *Political Science Quarterly*, vol. 98, no. 1 (1983): pp. 79–97. https://doi.org/10.2307/2150206.

22   Arms Control Association, "The Anti-Ballistic Missile (ABM) Treaty at a Glance," at https://www.armscontrol.org/factsheets/abmtreaty#:~:text=The%20United%20States%20and%20the%20Soviet%20Union%20negotiated%20the%20ABM,that%20the%20other%20might%20deploy.

23   The Guardian, "From the Archive, May 27, 1972: Nixon and Brezhnev Sign Historic Arms Treaty," *TheGuardian*, at https://www.theguardian.com/theguardian/2013/may/27/nuclear-arms-pact-russia-usa-1972. Accessed April 28, 2022.

24   Global Security, "SLBM Overview," at https://www.globalsecurity.org/wmd/world/russia/slbm-overview.htm. For a list of the Soviet SLBMs, see FAS, "Submarine Launched Ballistic Missiles," https://fas.org/nuke/guide/russia/slbm/index.html. See also W. Harriet Critchley, "Polar Deployment of Soviet Submarines," *International Journal*, vol. 39, no. 4 (1984): pp. 828–65, www.jstor.org/stable/40202298. Accessed April 28, 2021.

25   See Wikipedia, "Submarine-launched Ballistic Missile," at https://en.wikipedia.org/wiki/Submarine-launched_ballistic_missile.

26   Office of the Historian, Department of State, "Strategic Arms Limitations Talks/Treaty (SALT) I and II," at https://history.state.gov/milestones/1969-1976/salt.

27   This Day in History, "Jimmy Carter and Leonid Brezhnev Sign the SALT-II Nuclear Treaty," at https://www.history.com/this-day-in-history/carter-and-brezhnev-sign-the-salt-ii-treaty.

28   Office of the Historian, Department of State, "Strategic Arms Limitations Talks."

29   Leonid Brezhnev, "Excerpts From Speech by Brezhnev on Nuclear Arms Talks," *New York Times*, May 19, 1982, at https://www.nytimes.com/1982/05/19/world/excerpts-from-speech-by-brezhnev-on-nuclear-arms-talks.html. Accessed April 28, 2022.

Chapter 5

1   Among the many fine books on Chernobyl and nuclear power in the USSR on which I have drawn, see Svetlana Alexievich, *Voices from Chernobyl*, trans. Keith Gessen (Normal, IL and London: Dalkey Archive Press, 2005); David Marples, *The Social Impact of the Chernobyl Disaster* (New York: St. Martin's Press, 1988); and *Chernobyl and Nuclear Power in the USSR* (New York: St. Martin's Press, 1986); Sonja Schmid, *Producing Power* (Cambridge: MIT Press, 2015); Zhores Medvedev, *Legacy of Chernobyl* (New York: W. W. Norton and Co., 1992); Tatiana Kasperski, *Les politiques de la radioactivité: Tchernobyl et la mémoire nationale en Biélorussie contemporaine* (Paris: Pétra, 2020); and Serhii Plokhy, *Chernobyl: The History of a Nuclear Catastrophe* (New York: Basic Books, 2018). See also my *Red Atom* (2005).

2   Chernobyl'.by, "Ok"ekt "Ukrytie"," at http://www.chernobyl.by/shelter/33-sooruzhenie.html

3   Sergei Belyakov, *Liquidator: The Chernobyl Story* (Singapore: World Scientific, 2019); and Alekseivich, *Voices*.

4   Pavlo Fedykovych, "Inside Slavutych, the City Created by the Chernobyl Explosion," *CNN*, August 20, 2019, at https://edition.cnn.com/travel/article/chernobyl-slavutych-ukraine/index.html.

5   Anatoly Chernaev, "Notes from the Politburo Session about Chernobyl," July 3, 1986, at https://nsarchive.gwu.edu/document/19502-national-security-archive-doc-15-minutes-cc.

6   I draw heavily in this section Tatiana Kasperski, "From Legacy to Heritage: The Changing Political and Symbolic Status of Military Nuclear Waste in Russia," *Cahiers du Monde Russe*, vol. 60, nos. 2–3 (2019): pp. 517–38, and her as yet unpublished "Russian Reactors, Russian Radioactive Waste: Russian nuclear ambitions and waste legacies?"

7   Dmitriy Evlanov, "The Techa River: 50 Years of Radioactive Problems," in IAEA collection, at https://inis.iaea.org/collection/NCLCollectionStore/_Public/33/011/33011261.pdf; and "Scenario T," http://www-ns.iaea.org/downloads/rw/projects/emras-aquatic-techa.pdf. All five production reactors were shut down between 1987 and 1990, but too late to save the river.

8   Evlanov, "The Techa River."

9   Evlanov, "The Techa River."

10   Thomas Cochran, Robert Norris and Kristen Suokko, "Radioactive Contamination at Chelyabinsk-65, Russia," *Annual Review of Energy Environment*, vol. 18 (1993): pp. 513–16.

11   "Russia Consolidates Spent Nuclear Fuel from Research Facilities at
     Mayak," August 16, 2019, at http://russiannuclearsecurity.com/russia-
     consolidates-spent-nuclear-fuel-from-research-facilities-at-mayak.
     The Russian Nuclear Security Site is no longer available with the note
     "We evaluate the impact of these changes on our work and have to
     withdraw the content of this website until we conclude it is acceptable
     for us to continue sharing this information."

12   Andrey Ozharovsky, "Who Needs a Nuclear Power Station in
     Bashkortostan," *Bellona*, August 20, 2013, at https://bellona.org/news/
     nuclear-issues/nuclear-russia/2013-08-feature-who-needs-a-nuclear-
     power-plant-in-bashkortostan.

13   Ozharovsky, "Lithuania Shuts Down Soviet-era NPP, but Being a
     Nuclear-Free Nation Is Still under Question," *Bellona*, January 11, 2010,
     at https://bellona.org/news/nuclear-issues/nuclear-issues-in-ex-soviet-
     republics/2010-01-lithuania-shuts-down-soviet-era-npp-but-being-
     a-nuclear-free-nation-is-still-under-question; and "Chronology of
     Seminal Events Preceding the Declaration of Lithuania's Independence,"
     Lithuanian Quarterly Journal of Arts and Sciences, vol. 36, no. 2
     (Summer 1990), at http://www.lituanus.org/1990_2/90_2_07.htm.

14   Yuriy Shcherbak, *Chernobyl:A Documentary Story* (1988).

15   Jane Dawson, *Econationalism* (Durham: Duke, 1996); Marples, *The
     Social Impact of the Chernobyl Disaster.*

16   Andrei Stsiapanau, "Nuclear Energy in Transition. The Ignalina Nuclear
     Power Plant from Soviet under Lithuanian Rule," in S. Liubimau and B.
     Cope, eds., *Re-tooling Knowledge Infrastructures in a Post-nuclear Town*
     (Vilnius: Vilnius Academy of Arts Press, 2021), pp. 44–57.

17   Nick Meo, "Chernobyl's Arch: Sealing off a Radioactive Sarcophagus,"
     November 26, 2013, at http://www.bbc.co.uk/news/magazine-25086097

18   Abandoned Playgrounds, "The Crimean Atomic Station—The
     Unfinished Abandoned KaZantip Reactor," November 25, 2018,
     at http://www.abandonedplaygrounds.com/the-crimean-atomic-
     energy-station-the-unfinished-abandoned-kazantip-reaktor/; Nuclear
     Threat Initiative, "Radioactive Containers Found in Crimea," *NTI*,
     October 14, 2005, at http://www.nti.org/analysis/articles/radioactive-
     containers-found-crimea/; and "Obnaruzheny radioaktivnyye
     konteynery," *Krymskaua Pravda* (Simferopol), October 18, 2005, in
     Integrum Techno database, www.integrum.ru

Chapter 6

1   Rosatom, http://www.rosatom.ru/en/about/nuclear_industry/russian_
    nuclear_industry/

Notes

2   "VVER Reactor," at http://www.rosatom.ru/en/rosatom-group/
      engineering-and-construction/modern-reactors-of-russian-design/

3   European Parliament, "Commissioning of the Astravets Nuclear
      Power Plant without Implementation of All Safety Recommendations,"
      November 5, 2020, at https://www.europarl.europa.eu/doceo/
      document/P-9-2020-006039_EN.html; Charles Digges, "Russian
      Nuclear Officials Attempt to Bury Construction Mishap at
      Belarusian Nuclear Power Plant," August 4, 2016, at https://bellona.
      org/news/nuclear-issues/nuclear-issues-in-ex-soviet-republics/2016-
      08-russian-nuclear-officials-attempt-to-bury-construction-mishap-at-
      belarusian-nuclear-power-plant; and Nastassia Iaumen, "Astravetskaia
      AES Vachyma Zatrtymanai Zhurnalistki," *Novy Chas*, April 28, 2013, at
      https://novychas.by/hramadstva/astravieckaja_aes_vacyma_zatry.

4   Government of the Russian Federation, "General'naia Skhema
      Razmeshcheniia Ob"ektov Elektroenergetiki do 2020 Goda,
      Pasporiazhenie no. 215-r, February 22, 2008,"; and Kurchatov Institute,
      O Strategii Iadernoi Energetiki Rossii do 2050 goda (Moscow: KIAE,
      2012).

5   See for example Government of the Russian Federation, "O
      General'noi Skheme Razmeshcheniia Ob"ektov Elektroenergetiki do
      2020 Goda," Rasporiazhenie no. 215-r, February 22, 2008, at http://
      pravo.levonevsky.org/bazaru09/raspor/sbor05/text05919/index21.htm.

6   Stefan Guth, "Wachtürme unter Kränen. Zwangsarbeit in der post-
      stalinistischen UdSSR am Beispiel der Atomstadt Ševčenko/Aktau,
      1970," *Traverse. Zeitschrift für Geschichte*, vol. 24, no. 1 (2017):
      pp. 153–159. On the Soviet breeder reactor program, see Josephson,
      Red Atom, pp. 47–80.

7   http://www.anti-atom.ru/ab/node/1932. Unfortunately, this page is no
      longer working. But see for example MChS, "Intsident na Beloiarskoi
      AES (SSSR), Sviazannyi s Bol'shoi Tech'iu v Parogeneratore no 5
      Energobloka BN-600," at http://rb.mchs.gov.ru/mchs/radiation_
      accidents/m_other_accidents/1982_god/Incident_na_Belojarskoj_
      AJES_SSSR_svjaza; and Andrei Ozharovskii, "Belioiarskaia AES: ChP
      ko Dniu Atomshchik," *Bellona*, September 28, 2013, at http://bellona.
      ru/2013/09/28/beloyarskaya-aes-chp-ko-dnyu-atomshhika/, among
      dozens of other reports.

8   A. V. Trapeznikov et al., *Migratsiia Radionuklidov v Presnovodnykh
      Ekosistemakh*, 2 vols. (Ekaterinburg: Izdatel'stvo Ural'skogo
      Universiteta, 2007).

9   Kirill Bortnikov, "Edinstvennyi v Mire: Na Beloiarskoi AES Rabotaet
      Reaktor na Bystrykh Neitronakh," *Vesti*, December 22, 2015, https://
      www.vesti.ru/doc.html?id=2700947.

10 Tass, "Stoimost' Stroitelstva Reaktora na Bystrykh Neitronakh BN-800 Otsenivaetsia v 145.6 Mlrd Rublei," January 20, 2016, at http://www. atominfo.ru/newsm/t0547.htm

11 RIA Novosti, "Russia Planning 3 Advanced Fast-Breeder Reactors at Beloyarsk Nuclear Power Plant by 2030," July 23, 2014, at http:// atominfo.ru/en/news3/c0959.htm

12 RAS, "Experty Priznali Rossiiskii Proekt 'AES Budushchego' Ekonomicheski Vygodnym," June 7, 2017, at http://www.atomic-energy.ru/news/2017/06/07/76623; and RIA Novosti, "Konstruktor: Novyi Rossiiskii Blok AES BN-1200 Mozhno Vozvesti za 10 let," June 26, 2017, at https://ria.ru/atomtec/20170626/1497313701.html.

13 Atominfo, "Glavgosekspertiza Odobrila Stroitel'stvo Pervoi v Mire Plavuchei AES na Chukotke," January 1, 2018, at http://www.atominfo. ru/newsr/y0398.htm; Tass, "Obshchaia Stoimost' Pervoi v Mire Plavuchei AES Sostavit 30 Mlrd Rublei," April 10, 2016, at http:// www.atominfo.ru/newso/v0295.htm; and Russian Atomic Society, "V Peterburge Dosrochno Izgotovili Parogeneratory dlia Pervoi v Mire Plavuchei AES," February 11, 2009, at http://www.atomic-energy.ru/news/2009/02/11/2200

14 Kendall Bailes, "Technology and Legitimacy: Soviet Aviation and Stalinism in the 1930s," *Technology and Culture*, vol. 17 (1976): pp. 55–81; and Scott Palmer, *Dictatorship of the Air* (Cambridge: Cambridge University Press, 2006).

15 President of the Russian Federation, "Na Bortu Ledokol '50 Let Pobedy' Vladimir Putin Provel Zasedanie …," May 2, 2007, at http:// kremlin.ru/events/president/news/39116

16 V. P. Kuznetsov, V. V. Kushtan, A. A. Mirzoev, "Arkticheskii Vyzov Mirnogo Atoma," *Stroim Flot Sil'noi Strany*, vol. 3, no. 20 (2014): p. 43.

17 Nikolai Gorshkov, "Kursk Closure Leaves Questions Unanswered," *BBC*, July 31, 2002, at http://news.bbc.co.uk/2/hi/europe/2164783.stm

18 World Nuclear Organization, "Nuclear-Powered Ships," September 2021, at https://world-nuclear.org/information-library/non-power-nuclear-applications/transport/nuclear-powered-ships.aspx#:~:text=Seven%20Russian%20Alfa%2Dclass%20submarines,enriched%20in%20U%2DBe%20fuel.

19 Thomas Nilsen, "Sevmash Launches 'Krasnoyarsk' Nuclear Submarine," *Barents Observer*, July 30, 2021, at https://thebarentsobserver.com/en/security/2021/07/sevmash-launches-krasnoyarsk-nuclear-submarine.

20 "Russia Test Fires Submarine-launched Hypersonic Tsirkon Missile for First Time," *Reuters*, October 4, 2021, at https://www.reuters.com/world/europe/russia-test-fires-submarine-launched-hypersonic-tsirkon-missile-first-time-2021-10-04/.

21   NTI, "Russia. Missile," at https://www.nti.org/learn/countries/russia/delivery-systems/; and NTI, "Russia. Nuclear," at https://www.nti.org/learn/countries/russia/nuclear/.

22   Shannon Bugos, "U.S. Completes INF Treaty Withdrawal," September 2019, at https://www.armscontrol.org/act/2019-09/news/us-completes-inf-treaty-withdrawal.

23   Vladimir Isachenkov, "New Russian Policy Allows Use of Atomic Weapons against Non-Nuclear Strike," *Defense News*, June 2, 2020, at https://www.defensenews.com/global/europe/2020/06/02/new-russian-policy-allows-use-of-atomic-weapons-against-non-nuclear-strike/.

24   BBC, "Putin Warns of Tough Russian Action if West Crosses 'Red Line'," April 21, 2020, at https://www.bbc.com/news/world-europe-56828813.

25   Anton Lavrov and Aleksei Ramm, "'Poseidon' v Lodke," February 11, 2021, at https://iz.ru/1123160/anton-lavrov-aleksei-ramm/poseidon-v-lodke-submarinu-gotoviat-k-ispytaniiam-iadernykh-robotov, and Nick Walsh, "Satellite Images Show Huge Russian Military Buildup in the Arctic," *CNN*, April 5, 2021, at https://edition.cnn.com/2021/04/05/europe/russia-arctic-nato-military-intl-cmd/index.html.

26   Sherri Goodman and Katarina Kertysova, *The Nuclearisation of the Russian Arctic: New Reactors, New Risks* (London: European Leadership Network, June 2020).

27   Andrew Kramer, "Russia Confirms Radioactive Materials Were Involved in Deadly Blast," *New York Times*, August 10, 2019, at https://www.nytimes.com/2019/08/10/world/europe/russia-explosion-radiation.html; and Iu. V. Peshkov, "Ob Avariinom, Ekstremal'no Vysokom i Vysokom Zagriaznenii Okruzhaiushchei Sredy …" *RosGidroMet*, August 26, 2019, at http://www.meteorf.ru/product/infomaterials/91/19679/?sphrase_id=243600.

28   "Putin Rasskazal o Pogibshikh v Severodvinsk …" *RIA Novosti*, November 21, 2019, at https://web.archive.org/web/20191122050141/https://ria.ru/20191121/1561433888.html.

29   "Nuclear Weapons Stockpiles and Tests," https://en.wikipedia.org/wiki/Historical_nuclear_weapons_stockpiles_and_nuclear_tests_by_country.

30   Ministry of Energy of Ukraine, "Ukraine Calls Chernobyl Seizure by Russian Troops a Nuclear Terrorism and Asks IAEA to Immediately Appeal to NATO," March 3, 2022, at https://www.kmu.gov.ua/en/news/ukrayina-nazvala-fakt-zahoplennya-chaes-vijskami-rf-yadernim-terorizmom-ta-prosit-magate-nevidkladno-zvernutisya-do-nato

A BRIEF BIBLIOGRAPHIC NOTE

I am honored to have consulted the studies of a number of scholars whose works have been essential for understanding the history, politics, and culture of nuclear physics and technologies, peaceful and military, in the former Soviet Union and in Russia to this day. Zhores Medvedev's *Legacy of Chernobyl* (New York: W. W. Norton and Co., 1992) remains my favorite book on the subject. David Holloway's *Stalin and the Bomb; the Soviet Union and Atomic Energy, 1939–1956* (New Haven: Yale University Press, 1994) is required reading on the history and politics of the dawn of the Soviet nuclear age and the Soviet atomic bomb project. In *Producing Power* (Cambridge: MIT Press, 2015), Sonja Schmid opened Minsredmash and the Ministry of Electrification to crucial scrutiny. Kate Brown's *Plutopia* (New York: Oxford, 2013) indicates that the nature of the nuclear enterprise was quite similar in the United States and USSR. Tatiana Kasperski's *Les politiques de la radioactivité: Tchernobyl et la mémoire nationale en Biélorussie contemporaine* (Paris: Pétra, 2020) reveals the importance of memory politics in understanding the past and present of the Chernobyl disaster.

The literature on military technologies is not my area of expertise. But Robert Norris, Thomas Cochran, and their colleagues at the NRDC and FAS produced important works on nuclear weapons, proliferation and other subjects over forty years that have been crucial to my understandings of the Arms Race. Cochran's *Making the Russian Bomb: From Stalin to Yeltsin* (New York: Routledge, 1995) is but one of many contributions. To understand the vast sums spent on the arms race, see Stephen I. Schwartz, *Atomic Audit: The Costs and Consequences of U.S. Nuclear Weapons Since 1940* (Washington, DC: Brookings Institution, 1998). I have also consulted heavily the works of the Nuclear Threat Initiative, the Comprehensive Nuclear Test-Ban Treaty Organization, the Federation of American Scientists, and other organizations as noted in the footnotes.

Nobel Laureate Svetlana Alexievich's *Voices from Chernobyl* (2005) hauntingly captures the human dimension of the disaster and its consequences. David Marples wrote two early studies on Chernobyl and nuclear power, *The Social Impact of the Chernobyl Disaster* (New York: St. Martin's Press, 1988) and *Chernobyl and Nuclear Power in the USSR* (New York: St. Martin's Press, 1986), that set the stage for other research on the subject.

Finally, long ago, I wrote *Red Atom* (Pittsburgh: University of Pittsburgh Press, 2005) which pushed me in the direction of this *Russian Short*.

INDEX

Index

Index

nuclear fordism 64, 66–8
nuclear guinea pigs 32, 34–7
nuclear hubris 3, 5, 44, 99
Nuclear Non-Proliferation Treaty
    (NPT) 2
nuclear power 1–2, 5–7, 20–1, 42, 44,
    47, 49–50, 52, 54–5, 61–2,
    64, 68, 86, 94, 97–9, 107
nuclear-powered satellites 81
nuclear power plants (NPPs) 3,
    48, 54–5, 64–7, 81, 87,
    96–8, 101–2. *See also specific
    nuclear plants*
  floating 105–7, 112
  indigenes and 68–71
nuclear reactors. *See specific reactors*
nuclear renaissance 6, 97–8, 101–2, 105
nuclear strategy 42–3, 45
nuclear warheads 1–3, 5, 33, 44, 62, 74–6,
    78, 81, 109, 112, 114
nuclear waste/wasteland 1, 3, 6,
    34–5, 71, 81, 89–93. *See also*
    contamination
nuclear weapons 2–7, 9, 21–2, 25–6,
    32, 34–6, 41–5, 49–50, 54,
    57–9, 62–3, 68, 71, 74–6, 81,
    89–90, 109–11
Nunn-Lugar Cooperative Threat
    Reduction program 35

Obninsk 47–8
October Field 25–6, 93
Operation Alsos 24
Operation Plumbbob 36–7
Operation Project Paperclip 24
Oppenheimer, J. Robert 30

PAES. *See* floating nuclear power
    stations
particle accelerators 10, 12–13, 20,
    49, 56
peaceful atom 1, 3–5, 45, 47, 50–2,
    57–8, 69, 114, 123 n.3
peaceful nuclear explosions (PNEs) 5,
    34–5, 51, 60, 71–3, 81
perestroika 6, 88

Pevek 71, 105–6
Physicians for Social Responsibility 58
Podgorny, Nikolai 76
Poseidon (nuclear-armed torpedo) 112
postage stamps
    commemoration of twentieth
        anniversary of Atomic
        Icebreaker "Lenin" 101
    commemoration of victims in
        famine (1932–33) 15
    first-in-the world reactor to deliver
        power to civilian grid 48
    honor for contributions of
        Kurchatov 26
    honor for Flerov's 100th Birthday 24
    honor for Khariton's 100th Birthday 29
postwar atomic culture 50–2
pressurized water reactors (PWRs)
    2, 5, 54, 64, 66, 87, 93, 97,
    99–101, 103, 111
  VVER1200 PWR 2, 100–1
proliferation 42–3, 67
Putin, Vladimir 3, 6, 97, 101, 102–4,
    106–11, 113, 115

quantum mechanics 9, 19, 29–30, 50–1

Radiation Safety Service 73
radioactivity 3, 6, 10, 21–2, 25, 27, 32,
    36, 38–43, 58–60, 71, 83–7,
    91–6, 104, 112–13, 115
  radioactive contamination 39, 43,
      70, 84, 95
  radioactive fallout 3, 34, 59
  radioactive isotopes 13, 36, 56–7,
      91–2
  radioactive tracers in agriculture 51
  radioactive waste 33–5, 63, 72,
      89–92, 114–15
  soldiers exposed to 36
radionuclides 34, 60, 70, 72, 91, 93
RDS-1 31–2
RDS 2 (Joe 2) 32
RDS-3 32
RDS-4 32
RDS-6 (Joe 4) 32

138

Index

underground nuclear tests 35, 39, 60, 72
underwater nuclear tankers 107
The United Kingdom 33, 42, 60
The United Nations 4, 123 n.3
    Eisenhower's speech at 4, 47
    UN General Assembly 41, 43
The United States 1–3, 5, 21, 23, 25, 33–4, 36–8, 41–5, 50–2, 57, 59–62, 64, 67–8, 71, 74–6, 78–9, 81, 88, 100, 105–6, 108–10, 112, 114
    expenditures on weapons 75
    Project Plowshares 71
    US-Kazakhstan program 35
    US Manhattan Project 21, 23, 30, 58
    US Operation Crossroads 38
Ural Electrochemical Combine, Russia 33
uranium 1, 21–3, 25, 27–8, 30, 31–4, 40, 44, 51, 53–4, 67, 88, 99, 107
Uranium Committee (Academy of Sciences) 25
USSR 1, 3, 5–6, 9, 12, 16, 18, 22, 24, 30, 32–3, 35, 37, 41–3, 47, 50–2, 56–7, 59–62, 64, 73–7, 79–80, 84, 87–9, 93, 95–6, 98, 111, 118 n.1 (Intro)

indigenes and nuclear power in 68–71
and Irish General Assembly resolution 43
nuclear power stations in 65–7

Valter, Anton 12, 18–19
Vladivostok Summit, 1974 77
von Braun, Werner 24
VVER (Soviet PWR) 2, 64, 66, 81, 87, 98, 100–1. *See also* pressurized water reactors
VVR-M reactors 20

Warsaw Treaty 62, 80
Weapons of Mass Destruction (WMD) 35, 40, 44, 48, 76

Yeltsin, Boris 6, 90, 97

Zaporizhzhya NPP, Ukraine 115
Zarechnyi 104
ZATO (Closed Administrative-Territorial Formation) 4, 22, 25, 33, 35, 74, 81, 105, 108, 112, 120 n.3
Zelenyi Svit organization 93
Zhdanovshchina 29